Uncertainty in Risk Assessment

Uncertainty in Risk Assessment
The Representation and Treatment of Uncertainties by Probabilistic and Non-Probabilistic Methods

Terje Aven

University of Stavanger, Norway

Piero Baraldi

Politecnico di Milano, Italy

Roger Flage

University of Stavanger, Norway

Enrico Zio

Politecnico di Milano, Italy
Ecole Centrale Paris and Supelec, France

WILEY

This edition first published 2014
© 2014 John Wiley & Sons, Ltd

Registered office
John Wiley & Sons Ltd, The Atrium, Southern Gate, Chichester, West Sussex, PO19 8SQ, United Kingdom

For details of our global editorial offices, for customer services and for information about how to apply for permission to reuse the copyright material in this book please see our website at www.wiley.com.

Library of Congress Cataloging-in-Publication Data

Aven, Terje, author.
    Uncertainty in risk assessment : the representation and treatment of uncertainties by probabilistic and non-probabilistic methods / Terje Aven, Piero Baraldi, Roger Flage, Enrico Zio.
    pages cm
Includes bibliographical references and index.
    ISBN 978-1-118-48958-1 (hardback)
    1. Risk assessment–Statistical methods. 2.   Probabilities. I. Title.
    HD61.A947 2014
    338.5–dc23

                                                                              2013034152

A catalogue record for this book is available from the British Library.

ISBN: 978-1-118-48958-1

Set in 10/12 pt TimesLTStd-Roman by Thomson Digital, Noida, India

1   2014

Contents

Preface

The aim of this book is to critically present the state of knowledge on the treatment of uncertainties in risk assessment for practical decision-making situations concerning high-consequence technologies, for example, nuclear, oil and gas, transport, and so on, and the methods for the representation and characterization of such uncertainties. For more than 30 years, probabilistic frameworks and methods have been used as the basis for risk assessment and uncertainty analysis, but there is a growing concern, partly motivated by newly emerging risks like those related to security, that extensions and advancements are needed to effectively treat the different sources of uncertainty and related forms of information. Alternative approaches for representing uncertainty have been proposed, for example, those based on interval probability, possibility, and evidence theory. It is argued that these approaches provide a more adequate treatment of uncertainty in situations of poor knowledge of the phenomena and scenarios studied in the risk assessment. However, many questions concerning the foundations of these approaches and their use remain unanswered.

In this book, we present a critical review and discussion of methods for the representation and characterization of the uncertainties in risk assessment. Using examples, we demonstrate the applicability of the various methods and point to their strengths and weaknesses in relation to the situation addressed. Today, no authoritative guidance exists on when to use probability and when to use an alternative representation of uncertainty, and we hope that the present book can provide a platform for the development of such guidance. The areas of potential application of the theories and methods studied in the book are broad, ranging from engineering and medicine to environmental impacts and natural disasters, security, and financial risk management. Our main focus, however is, on engineering applications.

The topic of uncertainty representation and characterization is conceptually and mathematically challenging, and much of the existing literature in the field is not easily accessible to engineers and risk analysts. One aim of the present book is to provide a relatively comprehensive state of knowledge, with strong requirements for rigor and precision, while striving for readability by a broad audience of professionals in the field, including researchers and graduate students.

Readers will require some fundamental background in risk assessment, as well as basic knowledge of probability theory and statistics. The goal, however, has been to reduce the dependency on extensive prior knowledge, and key probabilistic and statistical concepts will be introduced and discussed thoroughly in the book.

It is with sincere appreciation that we thank all those who have contributed to the preparation of this book. In particular, we are grateful to Drs. Francesco Cadini, Michele Compare, Jan Terje Kvaløy, Giovanni Lonati, Irina Crenguza Popescu, Ortwin Renn, and Giovanna Ripamonti for contributing the research that has provided the material for many parts of the book, and to Andrea Prestigiacomo for his careful editing work. We also acknowledge the editing and production staff at Wiley for their careful and effective work.

Terje Aven
Roger Flage
Stavanger
Piero Baraldi
Milano
Enrico Zio
Paris

June 2013

Part I
INTRODUCTION

1

Introduction

Risk assessment is a methodological framework for determining the nature and extent of the risk associated with an activity. It comprises the following three main steps:

- Identification of relevant sources of risk (threats, hazards, opportunities)
- Cause and consequence analysis, including assessments of exposures and vulnerabilities
- Risk description.

Risk assessment is now widely used in the context of various types of activities as a tool to support decision making in the selection of appropriate protective and mitigating arrangements and measures, as well as in ensuring compliance with requirements set by, for example, regulatory agencies. The basis of risk assessment is the systematic use of analytical methods whose quantification is largely probability based. Common methods used to systematically analyze the causes and consequences of failure configurations and accident scenarios are fault trees and event trees, Markov models, and Bayesian belief networks; statistical methods are used to process the numerical data and make inferences. These modeling methods have been developed to gain knowledge about cause–effect relationships, express the strength of these relationships, characterize the remaining uncertainties, and describe, in quantitative or qualitative form, other properties relevant for risk management (IAEA, 1995; IEC, 1993). In short, risk assessments specify what is at stake, assess the uncertainties of relevant quantities, and produce a risk description which provides information useful for the decision-making process of risk management.

In this book we put the main focus on quantitative risk assessment (QRA), where risk is expressed using an adequate representation of the uncertainties involved. To further develop the methodological framework of risk assessment, we will need to explain in more detail what we mean by risk.

Uncertainty in Risk Assessment: The Representation and Treatment of Uncertainties by Probabilistic and Non-Probabilistic Methods, First Edition. Terje Aven, Piero Baraldi, Roger Flage and Enrico Zio.
© 2014 John Wiley & Sons, Ltd. Published 2014 by John Wiley & Sons, Ltd.

This introductory chapter is organized as follows. Following Section 1.1, which addresses the risk concept, we present in Section 1.2 the main features of probabilistic risk assessment (PRA), which is a QRA based on the use of probability to characterize and represent the uncertainties. Then, in Section 1.3, we discuss the use of risk assessment in decision-making contexts. Section 1.4 considers the issue of uncertainties in risk assessment, motivated by the thesis that if uncertainty cannot be properly treated in risk assessment, the risk assessment tool fails to perform as intended (Aven and Zio, 2011). This section is followed by a discussion on the main challenges of the probability-based approaches to risk assessment, and the associated uncertainty analysis. Alternative approaches for dealing with uncertainty are briefly discussed.

1.1   Risk

1.1.1   The concept of risk

In all generality, risk arises wherever there exists a potential source of damage or loss, that is, a hazard (threat), to a target, for example, people, industrial assets, or the environment. Under these conditions, safeguards are typically devised to prevent the occurrence of the hazardous conditions, and protection is put in place to counter and mitigate the associated undesired consequences. The presence of a hazard does not in itself suffice to define a condition of risk; indeed, inherent in the latter there is the uncertainty that the hazard translates from potential to actual damage, bypassing safeguards and protection. In synthesis, the notion of risk involves some kind of loss or damage that might be received by a target and the uncertainty of its transformation in actual loss or damage, see Figure 1.1. Schematically we can write (Kaplan and Garrick, 1981; Zio, 2007; Aven, 2012b)

$$\text{Risk} = \text{Hazards/Threats and Consequences (damage)} + \text{Uncertainty.} \qquad (1.1)$$

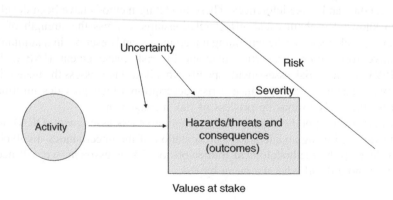

Figure 1.1   The concept of risk reflecting hazards/threats and consequences and associated uncertainties (what events will occur, and what the consequences will be).

Normally, the consequence dimension relates to some type of undesirable outcome (damage, loss, harm). Note that by centering the risk definition around undesirable outcomes, we need to define what is undesirable, and for whom. An outcome could be positive for some stakeholders and negative for others: discussing whether an outcome is classified in the right category may not be worth the effort, and most of the general definitions of risk today allow for both positive and negative outcomes (Aven and Renn, 2009).

Let A denote a hazard/threat, C the associated consequences, and U the uncertainties (will A occur, and what will C be?). The consequences relate to something that humans value (health, the environment, assets, etc.). Using these symbols we can write (1.1) as

$$\text{Risk} = (A, C, U), \tag{1.2}$$

or simply

$$\text{Risk} = (C, U), \tag{1.3}$$

where C in (C, U) expresses all consequences of the given activity, including the hazardous/threatful events A. These two risk representations are shown in Figure 1.2.

Obviously, the concept of risk cannot be limited to one particular measuring device (e.g., probability) if we seek a general risk concept. For the measure introduced, we have to explain precisely what it actually expresses. We also have to clarify the limitations with respect to its ability to measure the uncertainties: is there a need for a supplement to fully describe the risk? We will thoroughly discuss these issues throughout the book.

A concept closely related to risk is vulnerability (given the occurrence of an event A). Conceptually vulnerability is the same as risk, but conditional on the occurrence of an event A:

$$\text{Vulnerability} \mid A = \text{Consequences} + \text{Uncertainty} \mid \text{the occurrence of the event } A, \tag{1.4}$$

where the symbol $\mid$ indicates "given" or "conditional." For short we write

$$\text{Vulnerability} \mid A = (C, U \mid A). \tag{1.5}$$

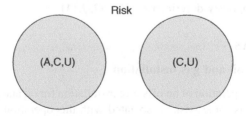

A: Events, C: Consequences, U: Uncertainty

Figure 1.2   The main components of the concept of risk used in this book.

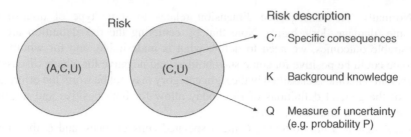

Figure 1.3   Illustration of how the risk description is derived from the concept of risk.

1.1.2   Describing/measuring risk

The risk concept has been defined above. However, this concept does not give us a tool for assessing and managing risk. For this purpose we must have a way of describing or measuring risk, and the issue is how.

As we have seen, risk has two main dimensions, consequences and uncertainty, and a risk description is obtained by specifying the consequences C and using a description (measure) of the uncertainty, Q. The most common tool is probability P, but others exist and these also will be given due attention in the book. Specifying the consequences means identifying a set of quantities of interest C' that represent the consequences C, for example, the number of fatalities.

Now, depending on the principles laid down for specifying C' and the choice of Q, we obtain different perspectives on how to describe/measure risk. As a general description of risk we can write

$$\text{Risk description} = (C', Q, K),$$
$$\text{(or, alternatively,} (A', C', Q, K)),$$

(1.6)

where K is the background knowledge (models and data used, assumptions made, etc.) that Q and the specification C' are based on, see Figure 1.3. On the basis of the relation between vulnerability and risk previously introduced, the vulnerability given an event A is analogously described by $(C', Q, K|A)$.

1.1.3   Examples

1.1.3.1   Offshore oil and gas installation

Consider the future operation of an offshore installation for oil and gas processing. We all agree that there is some "risk" associated with this operation. For example, fires and explosions could occur leading to fatalities, oil spills, economic losses, and so on. Today we do not know if these events will occur and what the specific consequences will be: we are faced with uncertainties and, thus, risk. Risk is two dimensional, comprising events and consequences, and associated uncertainties (i.e., the events and consequences being unknown, the occurrences of the events are not known and the consequences are not known).

When performing a risk assessment we describe and/or quantify risk, that is, we specify (C', Q, K). For this purpose we need quantities representing C' and a measure of uncertainty; for the latter, probability is introduced. Then, in the example discussed, C' is represented by the number of fatalities, $Q = P$, and the background knowledge K covers a number of assumptions that the assessment is based on, for example, related to the number of people working on the installation, as well as the models and data used for quantification of the accident probabilities and consequences. On this basis, several risk indices or metrics are defined, such as the expected number of fatalities (e. g., potential loss of lives, PLL, typically defined for a one-year period) and the fatal accident rate (FAR, associated with 100 million exposed hours), the probability that a specific person will be killed in an accident (individual risk, IR), and frequency–consequence $(f–n)$ curves expressing the expected number of accidents (frequency f) with at least n fatalities.

1.1.3.2   Health risk

Consider a person's life and focus on the condition of his/her health. Suppose that the person is 40 years old and we are concerned about the "health risk" for this person for a predetermined period of time or for the rest of his/her life. The consequences of interest in this case arise from "scenarios" of possible specific diseases (known or unknown types) and other illnesses, their times of development, and their effects on the person (will he/she die, suffer, etc.).

To describe risk in this case we introduce the frequentist probability p that the person gets a specific disease (interpreted as the fraction of persons that get the disease in an infinite population of "similar persons"), and use data from a sample of "similar persons" to infer an estimate p^* of p. The probability p can be considered a parameter of a binomial probability model.

For the consequent characterization, C', we look at the occurrence or not of a disease for the specific person considered, and the time of occurrence of the disease, if it occurs. In addition, we have introduced a probability model with a parameter p and this p also should be viewed as a quantity of interest C'. We seek to determine p, but there are uncertainties about p and we may use confidence intervals to describe this uncertainty, that is, to describe the stochastic variation in the data.

The uncertainty measure in this case is limited to frequentist probabilities. It is based on a traditional statistical approach. Alternatively, we could have used a Bayesian analysis based on subjective (judgmental, knowledge-based) probabilities P (we will return to the meaning of these probabilities in Chapter 2). The uncertainty description in this case may include a probability distribution of p, for example, expressed by the cumulative distribution function $F(p') = P(p \leq p')$. Using P to measure the uncertainties (i.e., $Q = P$), we obtain a risk description (C', P, K), where p is a part of C'. From the distribution $F(p')$ we can derive the unconditional probability $P(A)$ (more precisely, $P(A|K)$) of the event A that the person gets the disease, by conditioning on the true value of p (see also Section 2.4):

$$P(A) = \int P(A \mid p') \, dF(p') = \int p' \, dF(p'). \tag{1.7}$$

This probability is a subjective probability, based on the probability distribution of the frequentist probability p. We see that $P(A)$ is given by the center of gravity (the expected value) of the distribution F.

Alternatively, we could have made a direct subjective probability assignment for $P(A) = P(A \mid K)$, without introducing the probability model and the parameter p.

1.2   Probabilistic risk assessment

Since the mid-1970s, the framework of probability theory has been the basis for the analytic process of risk assessment (NRC, 1975); see the reviews by Rechard (1999, 2000). A probabilistic risk assessment (PRA) systematizes the knowledge and uncertainties about the phenomena studied: what are the possible hazards and threats, their causes and consequences? The knowledge and uncertainties are characterized and described using various probability-based metrics, as illustrated in Section 1.1.3; see also Jonkman, van Gelder, and Vrijling (2003) for a comprehensive overview of risk metrics (indices) for loss of life and economic damage. Additional examples will be provided in Chapter 3, in association with some of the detailed modeling and tools typical of PRA.

A total PRA for a system comprises the following stages:

1. *Identification of threats/hazards.* As a basis for this activity an analysis of the system is carried out in order to understand how the system works, so that departures from normal, successful operation can be identified. A first list of hazards/threats is normally identified based on this system analysis, as well as on experience from similar types of analyses, statistics, brainstorming activities, and specific tools such as failure mode and effect analysis (FMEA) and hazards and operability (HAZOP) studies.

2. *Cause analysis.* In cause analysis, we study the system to identify the conditions needed for the hazards/threats to occur. What are the causal factors? Several techniques exist for this purpose, from brainstorming sessions to the use of fault tree analyses and Bayesian networks.

3. *Consequence analysis.* For each identified hazard/threat, an analysis is carried out addressing the possible consequences the event can lead to. Consequence analysis deals to a large extent with the understanding of physical phenomena, for example, fires and explosions, and various types of models of the phenomena are used. These models may for instance be used for answering questions like: How will a fire develop? What will be the heat at various distances? What will the explosive pressure be in case an explosion takes place? And so on. Event tree analysis is a common method for analyzing the

scenarios that can develop in the different consequences. The number of steps in the sequence of events that form a scenario is mainly dependent on the number of protective barriers set up in the system to counteract the initiating event of that sequence. The aim of the consequence-reducing barriers is to prevent the initiating events from resulting in serious consequences. For each of these barriers, we can carry out failure analysis to study their reliability and effectiveness. Fault tree analysis is a technique often used for this purpose.

4. *Probabilistic analysis.* The previous stages of analysis provide a set of sequences of events (scenarios), which lead to different consequences. This specification of scenarios does not address the question of how likely the different scenarios and the associated consequences are. Some scenarios could be very serious, should they occur, but if the likelihood of their occurrence is low, they are not so critical. Using probability models to reflect variation in the phenomena studied and assigning probabilities for the occurrence of the various events identified and analyzed in steps 2 and 3, overall probability values and expected consequence values can be computed.

5. *Risk description.* Based on the cause analysis, consequence analysis, and probabilistic analysis, risk descriptions can be obtained using various metrics, for example, risk matrices showing the computed/assigned probability of a hazard/threat and the expected consequences given that this event has occurred, as well as IR, PLL, and FAR values.

6. *Risk evaluation.* The results of the risk analysis are compared to predefined criteria, for example, risk tolerability limits or risk acceptance criteria.

PRA methodology is nowadays used extensively in industries such as nuclear power generation (e.g., Vesely and Apostolakis, 1999; Apostolakis, 2004), offshore petroleum activities (e.g., Falck, Skramstad, and Berg, 2000; Vinnem, 2007), and air transport (e.g., Netjasov and Janic, 2008).

The current default approach to a comprehensive quantitative PRA is based on the so-called set of triplets definition of risk, introduced by Kaplan and Garrick (1981); see also Kaplan (1992, 1997). In this approach, risk is defined as the combination of possible scenarios s, resulting consequences x, and the associated likelihoods l. Loosely speaking: What can happen (go wrong)? How likely is it? What are the consequences? Within this conceptual framework, three main likelihood settings are often defined (Kaplan, 1997): repetitive situation with known frequency ($l = f$, where f is a known frequentist probability), unique situation ($l = p$, where p is a subjective probability), and repetitive situation with unknown frequency ($l = H(f)$, where H is a subjective probability distribution on an unknown/uncertain frequentist probability f). Of course, the first case is a special case of the third. The last-mentioned setting is typically dealt with using the so-called probability of frequency approach, where all potentially occurring events involved are assumed to have uncertain frequency probabilities of occurrence, and the epistemic uncertainties about the true values of frequency probabilities are described using subjective probabilities. For the sake of simplicity, in the following we will often use the short term "frequency" instead of "frequentist probability."

The probability of frequency approach is in line with the standard Bayesian approach (Aven, 2012a) as will be described below. It is also considered "the most general and by far the most powerful and useful idea" by Kaplan (1997, p. 409), and corresponds to the highest level of sophistication in the treatment of uncertainties in risk analysis according to the classification by Paté-Cornell (1996).

In this book, however, we adopt a broader perspective of risk by which the set of triplets is not *risk* per se but a *risk description*. In this view, the outcome of a risk assessment is a list of scenarios quantified in terms of probabilities and consequences, which collectively describe the risk. As we will thoroughly discuss throughout the book, this risk description will be shown to be more or less adequate for describing the risk and uncertainties in different situations.

Numerous textbooks deal with methods and models for PRA, for example, Andrews and Moss (2002), Aven (2008), Cox (2002), Vinnem (2007), Vose (2008), and Zio (2007, 2009). Some also deal specifically with foundational issues, in particular with the concepts of uncertainty and probability, for example, Aven (2012a), Bedford and Cooke (2001), and Singpurwalla (2006).

In spite of the maturity reached by the methodologies used in PRA, a number of new and improved methods have been developed in recent years to meet the needs of the analysis brought about by the increasing complexity of the systems and processes studied, and to respond to the introduction of new technological systems. Many of the methods introduced allow for increased levels of detail and precision in the modeling of phenomena and processes within an integrated framework of analysis covering physical phenomena, human and organizational factors, and software dynamics (e.g., Mohaghegh, Kazemi, and Mosleh, 2009). Other methods are devoted to the improved representation and assessment of risk and uncertainty. Examples of more recently developed methods are Bayesian belief networks, binary digit diagrams, multi-state reliability analysis, and advanced Monte Carlo simulation tools. For a summary and discussion of some of these models and techniques, see Bedford and Cooke (2001) and Zio (2009).

The probabilistic analysis underpinning PRA is based on one or the other of two alternative conceptual foundations: the traditional frequentist approach and the Bayesian approach (Bedford and Cooke, 2001; Aven, 2012a). The former is typically applied in situations in which there exists a large amount of relevant data; it is founded on well-known principles of statistical inference, the use of probability models, the interpretation of probabilities as relative frequencies, point estimates, confidence interval estimation, and hypothesis testing.

By contrast, the Bayesian approach is based on the concept of subjective (judgmental, knowledge-based) probabilities and is applied in situations in which there exists only a limited amount of data (e.g., Guikema and Paté-Cornell, 2004). The idea is to first establish probability models that adequately represent the aleatory uncertainties, that is, the inherent variability of the phenomena studied, such as the distribution of lifetimes of a type of system. The epistemic uncertainties, reflecting incomplete knowledge or lack of knowledge about the values of the parameters of the models, are then represented by prior subjective probability distributions. When new data on the phenomena studied becomes available, Bayes' formula is used to update the representation of the epistemic uncertainties in terms of the posterior distributions. Finally, the predictive distributions of the quantities of interest – the observables (e.g., the lifetime of new systems) – are derived by applying the law of total probability. The predictive distributions are epistemic statements, but they also reflect the inherent variability of the phenomena being studied, that is, the aleatory uncertainties.

1.3   Use of risk assessment: The risk management and decision-making context

Risk management can be defined as the coordinated activities to direct and control an organization with regard to risk (ISO, 2009). As illustrated in Figure 1.4, the main central steps of the risk management process are: establishment of the context, risk assessment, and risk treatment. Context here refer to the internal and external environment of the organization, the interface of these environments, the purpose of the risk management activity, and suitable risk criteria. Risk treatment is the process of modifying risk, which may involve avoiding, modifying, sharing or retaining risk (ISO, 2009).

Note that, according to ISO (2009), source (hazard/threat/opportunity) identification is not included as part of risk analysis. Many analysts and researchers do prefer, however, to include this element in the definition of the scope of risk analysis, in

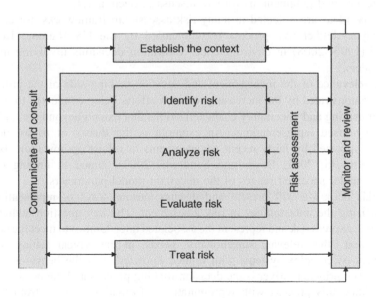

Figure 1.4   The risk management process (based on ISO, 2009).

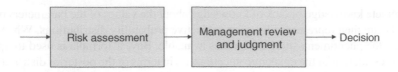

Figure 1.5   The leap (the management review and judgment) between risk assessment and the decision.

addition to cause and consequence analysis and risk description; see, for example, Modarres, Kamiskiy, and Krivtsov (1999) and Aven (2008).

There are different perspectives on how to *use* risk assessment and uncertainty analysis in risk management and decision making. Strict adherence to expected utility theory, cost–benefit analysis, stochastic optimization, and related theories would give clear recommendations on what is the optimal arrangement or measure. However, most risk researchers and risk analysts would see risk and uncertainty assessments as decision support tools, in the sense that the assessments inform the decision makers. The decision making is risk informed, not risk based (Apostolakis, 2004). In general, there is a significant leap from the assessments to the decision, see Figure 1.5. What this leap (often referred to as management review and judgment, or risk evaluation) comprises is a subject being discussed in the literature (e.g., Renn, 2005; Aven and Renn, 2010; Aven and Vinnem, 2007). The management review and judgment is about giving weight to the cautionary and precautionary policies and risk perception, as well as other concerns/attributes than risk and uncertainties. The scope and boundaries of risk and uncertainty assessments define to a large extent the content of the review and judgment, as we will discuss in Section 1.4.

Similar ideas are reflected in many risk assessment frameworks, for example, the analytic–deliberative process recommended by the US National Research Council (1996, 2008) in environmental restoration decisions involving multiple stakeholders.

The relevance of the management and decision-making side of the problem is further demonstrated by the increasing research efforts being conducted to integrate decision-making and uncertainty characterizations that extend beyond the traditional probability-based representations. An example is the theory of robust decision making, supported by fairly recent advancements in robust optimization (Ben-Tal and Nemirovski, 2002; Beyes and Sendhoff, 2007), aimed at finding optimal decisions under specified ranges of the uncertain model parameters.

In this book we will present and discuss various ways of representing and characterizing the uncertainties in risk assessment. The key question addressed is how to best express risk and represent the associated uncertainties to meet the decision makers' and other relevant stakeholders' needs, in the typical settings of risk assessment of complex systems with limited knowledge of the behavior of these systems. The principal driver is the decision-making process and the need to inform and facilitate this process with representative information derived from the risk assessment.

Hence, the book addresses, for example, the issue of how to present the results from risk and uncertainty assessments to decision makers, but not the decision making itself. We make a sharp distinction between risk/uncertainty representation and characterization on the one hand, and risk/uncertainty management and related decision making on the other.

1.4 Treatment of uncertainties in risk assessments

When speaking about uncertainties in risk assessments most analysts would think about the uncertainties related to parameters in probability models, such as the frequentist probability p in the second example in Section 1.1.3. Following the traditional statistical approach, the uncertainties are expressed using confidence intervals or according to the Bayesian approach, where subjective (judgmental, knowledge-based) probabilities are used to express the epistemic uncertainties about the parameters. This type of uncertainty analysis is an integrated part of risk assessment.

However, uncertainty analysis also exists independently of risk assessment (Morgan and Henrion, 1990). Formally, uncertainty analysis refers to the determination of the uncertainty associated with the results of an analysis that derives from uncertainty related to the input to the analysis (including the methods and models used in the analysis) (Helton *et al.*, 2006).

We may illustrate the ideas of uncertainty analysis by introducing a model $g(X)$, which depends on the input quantities X and on the function g. The quantity of interest, Z, is computed by using the model $g(X)$. The uncertainty analysis of Z requires an assessment of the uncertainties about X and their propagation through the model g to produce an assessment of the uncertainties concerning Z, see Figure 1.6. Uncertainty related to the model structure g, that is, uncertainty about the error $Z - g(X)$, is typically treated separately (Devooght, 1998; Zio and Apostolakis, 1996; Baraldi and Zio, 2010). In fact, while the impact of uncertainties associated with X has been widely investigated and many sophisticated methods have been developed to deal with it, research is still ongoing to obtain effective and agreed methods to handle the uncertainty related to the model structure (Parry and Drouin, 2009). For a broad sample of methods and ideas concerning model uncertainty, see Mosleh *et al.* (1994).

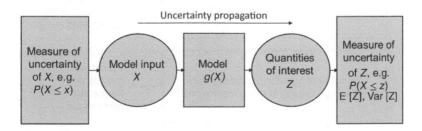

Figure 1.6 Basic features of uncertainty analysis (based on de Rocquigny, Devictor, and Tarantola (2008), Helton et al. *(2006), and Aven (2010a)).*

Overall frameworks for uncertainty analysis and management have been developed based on the elements Z and $g(X)$ as indicated above; see, for example, de Rocquigny, Devictor, and Tarantola (2008), Helton *et al.* (2006), and Aven (2010a).

These frameworks also applies to risk assessment. Then Z could for example be some high-level event of interest, such as a blowout in an offshore QRA setting or a core meltdown in a nuclear PRA setting, and X could be the set of low-level events which through various combinations could lead to the occurrence of the high-level event.

The quantities X and Z could also be frequentist probabilities representing fractions in a large (in theory, infinite) population of similar items, that is, parameters of probability models. Think of the frequentist probability p introduced in Section 1.1.3 that the person gets a specific disease. In this case, the assessment is consistent with the probability of frequency approach briefly outlined at the end of Section 1.1.3, which is based on the use of judgmental probabilities to express epistemic uncertainty about unknown frequentist probabilities (Kaplan and Garrick, 1981).

Finally, we add a note on sensitivity analysis, which is not the same as uncertainty analysis although they are closely linked. Sensitivity analysis indicates how sensitive the considered metrics are with respect to changes in basic input quantities (e.g., parameter values, assumptions and suppositions made) (Saltelli *et al.*, 2008; Cacuci and Ionescu-Bujor, 2004; Frey and Patil, 2002; Helton *et al.*, 2006). In an uncertainty analysis context, more specific definitions of sensitivity analysis have been suggested: for example, according to Helton *et al.* (2006), sensitivity analysis refers to the determination of the contributions of individual uncertain inputs to the analysis results.

In engineering risk assessments, a distinction is commonly made between aleatory and epistemic uncertainty (e.g., Apostolakis, 1990; Helton and Burmaster, 1996) as mentioned above in relation to the Bayesian approach, see Figure 1.7. Aleatory uncertainty refers to variation in populations, and epistemic uncertainty to lack of knowledge about phenomena. The latter usually translates into uncertainty about the parameters of a probability model used to describe variation. Whereas epistemic uncertainty can be reduced, aleatory uncertainty cannot, and it is sometimes called irreducible uncertainty (Helton and Burmaster, 1996). The aleatory uncertainty concept

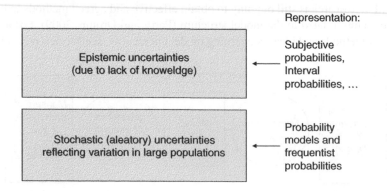

Figure 1.7   Illustration of the two types of uncertainties, aleatory and epistemic, and how they are represented.

is variably referred to as stochastic uncertainty (e.g., Helton and Burmaster, 1996), randomness, (random) variation (e.g., Aven, 2012a) or (random) variability (e.g., Baudrit, Dubois, and Guyonnet, 2006). It requires a large (in theory, infinite) population of "similar" but not identical units (realizations) to become operational, for example, a population of mass-produced units, a population of human beings or a sequence of dice throws. The epistemic uncertainty concept is variously referred to as subjective uncertainty (e.g., Helton and Burmaster, 1996), (partial) lack of knowledge (e.g., Aven, 2012a), or partial ignorance (e.g., Ferson and Ginzburg, 1996).

Returning to the health risk example of Section 1.1.3, the binomial probability model and the frequentist probability p represent the aleatory uncertainty, the variation in populations, whereas the subjective probability distribution $H(p')$ describes the epistemic uncertainties, reflecting the assessor's judgment about the true value of p.

1.5   Challenges: Discussion

In all the examples above we have used probabilities to describe risk. However, there are a growing number of researchers and analysts who find the probability-based approaches for assessing risk and uncertainties to be too narrow (see, e.g., Renn, 1998; Aven and Zio, 2011; Aven, 2011b). The argumentation follows different lines of thinking. One main point raised is that risk is more than some analysts' subjective probabilities, which may be based on strong assumptions and lead to poor predictions of Z. One may assign a low probability of health problems caused by the use of some new chemicals, but these probabilities could produce poor predictions of the actual number of people that experience such problems. Or one may assign a probability of fatalities occurring on an offshore installation based on the assumption that the installation structure will withstand a certain accidental collision energy load while, in reality, the structure could fail at a lower load level: the assigned probability did not reflect this uncertainty.

While probabilities can always be assigned under the subjective probability approach, the origin and amount of information supporting the assignments are not reflected by the numbers produced. One may, for example, subjectively assess that two different events have probabilities equal to, say, 0.7, but in one case the assignment is supported by a substantial amount of relevant data, whereas in the other by effectively no data at all. This is the main argument in the critique of the probability-based approach to risk and uncertainty.

Another important argument relates to the decision setting. In a risk assessment context there are often many stakeholders and they may not be satisfied with a probability-based assessment expressing the subjective judgments of the analysis group; a broader risk description is sought.

Probability models constitute a pillar of the probabilistic approach, an essential tool for assessing uncertainties and drawing useful insights (Helton, 1994; Winkler, 1996). The probability models coherently and mechanically facilitate the updating of probabilities. However, for many types of applications these models cannot be justified, the consequence being that the probability-based approach to risk and uncertainty becomes difficult to implement. A probability model presumes some sort of model stability; populations of similar units need to be constructed. But such stability is often not fulfilled (Bergman, 2009; Aven and Kvaløy, 2002). Consider the definition of a chance. In the case of a die, we would establish a probability model expressing that the distribution of outcomes is given by $(p_1, p_2, \ldots p_6)$, where p_i is the chance of outcome i, interpreted as the fraction of throws resulting in outcome i. However, in a risk assessment context the situations are often unique and the establishment of chances means the construction of fictional populations of non-existing similar situations. Then, chances and probability models in general cannot be easily defined as in the die-tossing example; in many cases they cannot be meaningfully defined at all. For example, it makes no sense to define a chance (frequentist probability) of a terrorist attack (Aven and Renn, 2010, p. 80). In other cases the conclusion may not be so obvious. For example, the chance of an explosion in a process plant may be introduced in a risk assessment, although the underlying population of infinite similar situations is somewhat difficult to describe.

1.5.1 Examples

1.5.1.1 LNG plant risk

An LNG (Liquefied Natural Gas) plant is planned and the operator would like to locate the plant no more than a few hundred meters from a residential area (Vinnem, 2010). Risk assessments are performed demonstrating that the risk is acceptable according to some predefined risk acceptance criteria. Risk is expressed using computed probabilities and expected values. Both IR numbers and f–n curves are presented. However, the assessments and the associated risk management are subject to strong criticism. The neighbors as well as many independent experts do not find the risk characterization sufficiently informative to support the decision making on the location and design of the plant. Sensitivity analyses are lacking, as well as reflections on uncertainties. The risk figures produced are based on a number of critical assumptions, but these assumptions are neither integrated into the risk characterization presented, nor communicated by the operator.

This type of risk assessment has been conducted for many years (Kolluru et al., 1996) but the problems seem to be the same. Still, a narrow risk characterization is provided. One reason may be the fact that acknowledging uncertainty can weaken the authority of the decision maker and agency, by creating an image of being unknowledgeable (Tickner and Kriebel, 2006). We will not, however, discuss this any further here; the point we would like to make is simply that these problems are also a result of the perspective adopted for the risk assessments. The conventional view, supported by most authoritative guides on risk assessments (e.g., Bedford and Cooke, 2001; Vose, 2008), is to a large extent pleased with the "narrow" probabilistic-based assessments as in the LNG case. Based on this view, the assessments may be criticized for too few and too restricted sensitivity and uncertainty analyses, but not for systematic hiding or camouflaging uncertainties.

1.5.1.2   Terrorism risk

For many of the currently emerging risks, the need for an extended approach is even more urgent. Take for example terrorism risk. Here the risk assessments often focus on estimating the probability of an attack, P(attack). But what is the meaning of such a probability? The conventional assessment is based on probability models, so there is a need for the definition of a frequentist probability or chance $p = P$(attack). But such interpretations are meaningless, as mentioned above (Aven and Renn, 2010, p. 80). To define such a probability (chance) we need to construct an infinite population of similar attack situations. The proportion of successes equals the probability of an attack. But it makes no sense to define a large set of "identical," independent attack situations, where some aspects (e.g., related to the potential attackers and the political context) are fixed and others (e.g., the attackers' motivation) are subject to variation. Say that the attack success rate is 10%. Then, in 1000 situations, with the attackers and the political context specified, the attackers will successfully attack in about 100 cases. In these situations the attackers are motivated, but not in the remaining ones. Motivation for an attack in one situation does not affect the motivation in another. For independent repeated random situations such "experiments" are meaningful, but not in unique cases like this. Still, many researchers and analysts work within such a risk assessment framework.

1.5.2   Alternatives to the probability-based approaches to risk and uncertainty assessment

Based on the above critiques, it is not surprising that alternative approaches for representing and describing risk and uncertainties have been suggested; see, for example, the special issues on imprecision in the *Journal of Statistical Theory and Practice* (JSTP, 2009), on uncertainty in engineering risk and reliability in the *Journal of Risk and Reliability* (JRR, 2010), and on alternative representations of epistemic uncertainty in *Reliability Engineering and System Safety* (RESS, 2004). Four main categories are (Dubois, 2010; Aven and Zio, 2011):

a. Probability-bound analysis, combining probability analysis and interval analysis (Ferson and Ginzburg, 1996). Interval analysis is used for those components whose aleatory uncertainties cannot be accurately estimated; for the other components, traditional probabilistic analysis is carried out.

b. Imprecise probability, after Walley (1991), and the robust statistics area (Berger, 1994) (see also Coolen, Augustin, and Troffaes, 2010; Klir, 2004), which encompasses probability-bound analysis, and certain aspects of evidence and possibility theory as special cases.

c. Evidence theory (or belief function theory), as proposed by Dempster (1967) and Shafer (1976), and the closely linked theory of random sets (Nguyen, 2006).

d. Possibility theory (Dubois and Prade, 1988, 2009; Dubois, 2006), which is formally a special case of the imprecise probability and random set theories.

We will thoroughly review and discuss these approaches in this book, as well as attempts made to combine different approaches, for example, probabilistic analysis and possibility theory, where the uncertainties of some parameters are represented by probability distributions and the uncertainties of the remaining parameters by means of possibility distributions; see Baudrit, Dubois, and Guyonnet (2006) and the applications in Baraldi and Zio (2008), Flage *et al.* (2013), and Helton, Johnson, and Oberkampf (2004).

All these approaches and methods produce epistemic-based uncertainty descriptions and in particular intervals, but they have not been broadly accepted by the risk assessment community. Much effort has been made in this area, but there are still many open questions related to the foundation of these approaches and their use in risk and uncertainty decision making; see the discussions in, for example, Cooke (2004), Smets (1994), Lindley (2000), Aven (2011a), Bernardo and Smith (1994), and North (2010). Many risk researchers and risk analysts are skeptical about the use of the alternative approaches (such as those of the four categories (a)–(d) mentioned above) for the representation and treatment of uncertainty in risk assessment for decision making, and some also argue intensively against them; see, for example, North (2010, p. 380).

However, as argued above, the probability-based approach does not solve the problems raised. The decision basis cannot be restricted to assigned probabilities: there is a need to go beyond the traditional Bayesian approach; a broader perspective and framework is required. The present book is based on such a conviction. As a matter of fact, no comprehensive authoritative guidance exists today on when to use probability and when to use an alternative representation of uncertainty. The main challenge is to define the conditions when probability is the appropriate representation of uncertainty. An argument often propounded is that probability is the appropriate representation of uncertainty only when a sufficient amount of data exists on which to base the probability (distribution) in question. But it is not obvious how to make such a prescription operational (Flage, 2010). Consider the representation of uncertainty about the parameter(s) of a probability model. If a large enough amount of data exists, there would be no uncertainty about the parameter(s) and hence no need for a representation of such uncertainty. When is there enough data to justify probability, but not enough to accurately specify the true value of the parameter in question and, thus, make probability, as an epistemic concept, superfluous?

Other approaches for representing the uncertainties have also been suggested. One example is the approach based on the maximum entropy principle. This approach does not require the specification of the whole probability distribution but only of some of its features, for example, the mean and variance; then, a mathematical procedure is applied to obtain the distribution characterized by the specified features and, in a certain sense, minimum information beyond that, see Bedford and Cooke (2001). Another approach relevant in this context is probabilistic inference with uncertain and partial evidence, developed by Groen and Mosleh (2005) as a generalization of Bayes' theorem.

Taking a broader view, we may identify different directions of development with respect to (alternative) representations of uncertainty in risk analysis (Flage, 2010).

One direction, as suggested by Lindley (2006) and O'Hagan and Oakley (2004), is to retain probability as the sole representation of uncertainty and to focus on improving the measurement procedures for probability. Another direction is that resulting in a semi-quantitative approach, where quantitative risk metrics are supported by qualitative assessments of the strength of background knowledge of these metrics (Aven, 2013). Such an approach presupposes an acknowledgment and belief that the full scope of risk and uncertainty cannot be transformed into a mathematical formula, using probability or any other measure of uncertainty. Numbers can be generated, but these alone would not serve the purpose of the risk assessment, to reveal and describe the risks and uncertainties. Furthermore, a duality in terms of interpretation, like that which affects probability (limiting relative frequency vs. degree of belief), also affects possibility theory (degree of compatibility or ease vs. lower and upper probability) and the theory of belief functions (degree of belief per se vs. lower and upper probability) (Flage, 2010). Another direction thus consists of assessing the appropriateness of lower and upper probability vs. other interpretations for risk analysis purposes, and then developing a proper foundational basis like those mentioned above in the Bayesian setting. Finally, there is the "unifying" approach, based on the combination of different representations. These development directions are further studied in Chapter 7.

1.5.3   The way ahead

When considering the methods for representing and characterizing uncertainties in risk assessment, two main concerns need to be balanced:

1. Knowledge should, as far as possible, be "inter-subjective" in the sense that the representation corresponds to "documented and approved" information and knowledge ("evidence"); the methods and models used to treat this knowledge should not add information that is not there, nor ignore information that is there.

2. Analysts' judgments ("degrees of belief") should be clearly reflected ("judgments").

The former concern can make the pure Bayesian approach difficult to apply in certain instances: when scarce information and little knowledge are available, introducing analysts' subjective probability distributions may be unjustifiable since this leads to building a structure in the analysis that is not present in the information and knowledge. For example, if an expert states his or her uncertainty assessment on a parameter value in terms of a range of possible values, this does not justify the allocation of a specific probability distribution function (e.g., the uniform probability distribution) onto the range. In this view, it might be said that a more defense-in-depth (bounding) representation of the information and knowledge available would be one which leaves the analysis open to all possible probability distribution structures on the assessed range, without imposing one in particular and without excluding any, thus providing results which bound all possible distributions.

At the same time, the representation framework should also take into account the second concern above, that is, allow for the transparent inclusion of preferential assignments by the experts (analysts) who wish to express that, according to their beliefs, some values are more or less likely than others. The Bayesian approach is the proper framework for such assignments.

From the point of view of the quantitative modeling of uncertainty in risk assessment, two topical issues are the proper handling of dependencies among uncertain parameters, and of model uncertainties. No matter what modeling paradigm is adopted, it is critical that the meaning of the various concepts be clarified. Without such clarification it is impossible to build a scientifically based risk assessment. In complex situations, when the propagation is based on many parameters, strong assumptions may be required to be able to carry out the analysis in practice. The risk analysts may acknowledge a degree of dependency, but the analysis may not be able to describe it in an adequate way. The derived uncertainty representations must be understood and communicated as measures conditional on this constraint. In practice it is the main task of the analysts to seek simple representations of the system performance, and by smart modeling it is often possible to obtain independence. The models used are also included in the background knowledge of epistemic-based uncertainty representations. We seek accurate models, but at the same time simple models. The choice of the right model cannot be seen in isolation from the purpose of the risk assessments.

Acknowledging risk and uncertainty assessments as decision support tools requires that the meaning and practical interpretation of the quantities computed are presented and communicated in an understandable format to the decision makers. The format must allow for meaningful comparisons to numerical safety criteria, if defined, for manipulation (e.g., by screening, bounding, and/or sensitivity analyses) and for communication in deliberation processes. This issue has been addressed by many researchers in the scientific literature; see, for example, the recent discussions in Aven (2010b), Dubois (2010), and Dubois and Guyonnet (2011) as well as Renn (2008). However, there are still many questions that remain to be answered, for example, concerning the type of information the decision maker needs in order to be risk informed, as the debate between Aven (2010b) and Dubois (2010) reveals. We will address this issue in detail in the coming chapters.

In the coming chapters, we will perform a detailed review and discussion of the most common and relevant approaches and methods for representing and characterizing uncertainties in risk assessment. The review and discussion are based on the perspective and framework of risk assessment and uncertainty analysis introduced in this chapter. This perspective and framework extend beyond the Bayesian approach. We have argued that a full risk–uncertainty description is more than subjective probabilities. Risk is about hazards/threats, their consequences, and the associated uncertainties, and to assess risk various tools can be used to measure the uncertainties. In the coming chapters, we will look closer at the most important of these tools, to provide the reader with an improved basis for selecting appropriate approaches and methods for representing and characterizing uncertainties in risk assessment.

References – Part I

Andrews, J.D. and Moss, T.R. (2002) *Reliability and Risk Assessment*, 2nd edn, Professional Engineering Publishing, London.

Apostolakis, G. (1990) The concept of probability in safety assessments of technological systems. *Science*, **250** (4986), 1359–1364.

Apostolakis, G.E. (2004) How useful is quantitative risk assessment? *Risk Analysis*, **24**, 515–520.

Aven, T. (2008) *Risk Analysis*, John Wiley & Sons, Ltd, Chichester.

Aven, T. (2010a) Some reflections on uncertainty analysis and management. *Reliability Engineering and System Safety*, **95**, 195–201.

Aven, T. (2010b) On the need for restricting the probabilistic analysis in risk assessments to variability. *Risk Analysis*, **30** (3), 354–360. With discussion on pp. 361–384.

Aven, T. (2011a) On the interpretations of alternative uncertainty representations in a reliability and risk analysis context. *Reliability Engineering and System Safety*, **96**, 353–360.

Aven, T. (2011b) Selective critique of risk assessments with recommendations for improving methodology and practice. *Reliability Engineering and System Safety*, **5**, 509–514.

Aven, T. (2012a) *Foundations of Risk Analysis*, 2nd edn, John Wiley & Sons, Ltd, Chichester.

Aven, T. (2012b) The risk concept: historical and recent development trends. *Reliability Engineering and System Safety*, **99**, 33–44.

Aven, T. (2013) Practical implications of the new risk perspectives. *Reliability Engineering and System Safety*, **115**, 136–145.

Aven, T. and Kvaløy, J.T. (2002) Implementing the Bayesian paradigm in risk analysis. *Reliability Engineering and System Safety*, **78**, 195–201.

Aven, T. and Renn, O. (2009) On risk defined as an event where the outcome is uncertain. *Journal of Risk Research*, **12**, 1–11.

Aven, T. and Renn, O. (2010) *Risk Management and Risk Governance*, Springer Verlag, London.

Aven, T. and Vinnem, J.E. (2007) *Risk Management, with Applications from the Offshore Oil and Gas Industry*, Springer Verlag, New York.

Aven, T. and Zio, E. (2011) Some considerations on the treatment of uncertainties in risk assessment for practical decision-making. *Reliability Engineering and System Safety*, **96**, 64–74.

Uncertainty in Risk Assessment: The Representation and Treatment of Uncertainties by Probabilistic and Non-Probabilistic Methods, First Edition. Terje Aven, Piero Baraldi, Roger Flage and Enrico Zio.
© 2014 John Wiley & Sons, Ltd. Published 2014 by John Wiley & Sons, Ltd.

Baraldi, P. and Zio, E. (2008) A combined Monte Carlo and possibilistic approach to uncertainty propagation in event tree analysis. *Risk Analysis*, **28** (5), 1309–1325.

Baraldi, P. and Zio, E. (2010) A comparison between probabilistic and Dempster–Shafer theory approaches to model uncertainty analysis in the performance assessment of radioactive waste repositories. *Risk Analysis*, **30** (7), 1139–1156.

Baudrit, C., Dubois, D., and Guyonnet, D. (2006) Joint propagation of probabilistic and possibilistic information in risk assessment. *IEEE Transactions on Fuzzy Systems*, **14**, 593–608.

Bedford, T. and Cooke, R. (2001) Probabilistic Risk Analysis: *Foundations and Methods*, Cambridge University Press, Cambridge.

Ben-Tal, A. and Nemirovski, A. (2002) Robust optimization – methods and applications. *Mathematical Programming*, **92** (3), 453–480.

Berger, J. (1994) An overview of robust Bayesian analysis. *Test*, **3**, 5–124.

Bergman, B. (2009) Conceptualistic pragmatism: a framework for Bayesian analysis? *IIE Transactions*, **41**, 86–93.

Bernardo, J.M. and Smith, A.F.M. (1994) *Bayesian Theory*, John Wiley & Sons, Ltd, Chichester.

Beyes, H.G. and Sendhoff, B. (2007) Robust optimization – a comprehensive survey. *Computer Methods in Applied Mechanics and Engineering*, **196** (33–34), 3190–3218.

Cacuci, D.G. and Ionescu-Bujor, M.A. (2004) Comparative review of sensitivity and uncertainty analysis of large-scale systems – II: statistical methods. *Nuclear Science and Engineering*, **147** (3), 204–217.

Cooke, R. (2004) The anatomy of the squizzel: the role of operational definitions in representing uncertainty. *Reliability Engineering and System Safety*, **85**, 313–319.

Coolen, F., Augustin, C.T., and Troffaes, M.C.M. (2010) Imprecise probability, in *International Encyclopedia of Statistical Science* (ed. M. Lovric), Springer Verlag, Berlin, pp. 645–648.

Cox, L.A. (2002) *Risk Analysis: Foundations, Models, and Methods*, Kluwer Academic, Boston, MA.

de Rocquigny, E., Devictor, N., and Tarantola, S. (eds.) (2008) *Uncertainty in Industrial Practice: A Guide to Quantitative Uncertainty Management*, John Wiley & Sons, Inc., Hoboken, NJ.

Dempster, A.P. (1967) Upper and lower probabilities induced by a multivalued mapping. *Annals of Mathematical Statistics*, **38**, 325–339.

Devooght, J. (1998) Model uncertainty and model inaccuracy. *Reliability Engineering and System Safety*, **59**, 171–185.

Dubois, D. (2006) Possibility theory and statistical reasoning. *Computational Statistics & Data Analysis*, **51**, 47–69.

Dubois, D. (2010) Representation, propagation and decision issues in risk analysis under incomplete probabilistic information. *Risk Analysis*, **30**, 361–368.

Dubois, D. and Guyonnet, D. (2011) Risk-informed decision-making in the presence of epistemic uncertainty. *International Journal of General Systems*, **40**, 145–167.

Dubois, D. and Prade, H. (1988) *Possibility Theory*, Plenum Press, New York.

Dubois, D. and Prade, H. (2009) Formal representations of uncertainty, in *Decision-Making Process: Concepts and Methods* (eds. D. Bouyssou, D. Dubois, M. Pirlot, and H. Prade), ISTE, London, pp. 85–156.

Falck, A., Skramstad, E., and Berg, M. (2000) Use of QRA for decision support in the design of an offshore oil production installation. *Journal of Hazardous Materials*, **71**, 179–192.

Ferson, S. and Ginzburg, L.R. (1996) Different methods are needed to propagate ignorance and variability. *Reliability Engineering and System Safety*, **54**, 133–144.

Flage, R. (2010) Contributions to the treatment of uncertainty in risk assessment and management. PhD thesis No. 100, University of Stavanger.

Flage, R., Baraldi, P., Ameruso, F. *et al.* (2010) *Handling Epistemic Uncertainties in Fault Tree Analysis by Probabilistic and Possibilistic Approaches*, ESREL 2009, Prague, 7–10 September 2009. CRC Press, London, pp. 1761–1768.

Flage, R., Baraldi, P., Zio, E., and Aven, T. (2013) Probabilistic and possibilistic treatment of epistemic uncertainties in fault tree analysis. *Risk Analysis*, **33** (1), 121–133.

Frey, H.C. and Patil, S.R. (2002) Identification and review of sensitivity analysis methods. *Risk Analysis*, **22** (3), 553–578.

Groen, F.J. and Mosleh, A. (2005) Foundations of probabilistic inference with uncertain evidence. *International Journal of Approximate Reasoning*, **39**, 49–83.

Guikema, S.D. and Paté-Cornell, M.E. (2004) Bayesian analysis of launch vehicle success rates. *Journal of Spacecraft and Rockets*, **41** (1), 93–102.

Helton, J.C. (1994) Treatment of uncertainty in performance assessments for complex systems. *Risk Analysis*, **14**, 483–511.

Helton, J.C. and Burmaster, D.E. (1996) Guest editorial: treatment of aleatory and epistemic uncertainty in performance assessments for complex systems. *Reliability Engineering and System Safety*, **54**, 91–94.

Helton, J.C., Johnson, J.D., and Oberkampf, W.L. (2004) An exploration of alternative approaches to the representation of uncertainty in model predictions. *Reliability Engineering and System Safety*, **85** (1–3), 39–71.

Helton, J.C., Johnson, J.D., Sallaberry, C.J., and Storlie, C.B. (2006) Survey of sampling-based methods for uncertainty and sensitivity analysis. *Reliability Engineering and System Safety*, **91**, 1175–1209.

IAEA (1995) Guidelines for Integrated Risk Assessment and Management in Large Industrial Areas, Technical Document: IAEA–TECDOC PGVI–CIJV, International Atomic Energy Agency, Vienna.

IEC (1993) Guidelines for Risk Analysis of Technological Systems, Report IEC–CD (Sec) 381 issued by Technical Committee QMS/23, European Community, Brussels.

ISO (2009) ISO 31000:2009, Risk management—Principles and guidelines.

Jonkman, S.N., van Gelder, P.H.A.J.M., and Vrijling, J.K. (2003) An overview of quantitative risk measures for loss of life and economic damage. *Journal of Hazardous Materials*, **99** (1), 1–30.

JRR (2010) Special issue on uncertainty in engineering risk and reliability. *Journal of Risk and Reliability*, **224** (4) (eds. F.P.A. Coolen, M. Oberguggenberger, and M. Troffaes).

JSTP (2009) Special issue on imprecision. *Journal of Statistical Theory and Practice*, **3** (1).

Kaplan, S. (1997) The words of risk analysis. *Risk Analysis*, **17**, 407–417.

Kaplan, S. (1992) Formalism for handling phonological uncertainties: the concepts of probability, frequency, variability, and probability of frequency. *Nuclear Technology*, **102**, 137–142.

Kaplan, S. and Garrick, B.J. (1981) On the quantitative definition of risk. *Risk Analysis*, **1**, 11–27.

Klir, G.J. (2004) Generalized information theory: aims, results, and open problems. *Reliability Engineering and System Safety*, **85**, 21–38.

Kolluru, R., Bartell, S., Pitblado, R., and Stricoff, S. (eds.) (1996) *Risk Assessments and Management Handbook*, McGraw-Hill, New York.

Lindley, D.V. (2000) The philosophy of statistics. *The Statistician*, **49** (3), 293–337.

Lindley, D.V. (2006) *Understanding Uncertainty*, John Wiley & Sons, Inc., Hoboken, NJ.

Modarres, M., Kamiskiy, M., and Krivtsov, V. (1999) *Reliability Engineering and Risk Analysis*, CRC Press, Boca Raton, FL.

Mohaghegh, Z., Kazemi, R., and Mosleh, A. (2009) Incorporating organizational factors into Probabilistic Risk Assessment (PRA) of complex socio-technical systems: a hybrid technique formalization. *Reliability Engineering and System Safety*, **94**, 1000–1018.

Morgan, M.G. and Henrion, M. (1990) Uncertainty, in *A Guide to Dealing with Uncertainty in Quantitative Risk and Policy Analysis*, Cambridge University Press, Cambridge.

Mosleh, A., Siu, N., Smidts, C., and Lui, C. (eds.) (1994) Model uncertainty: its characterization and quantification. Report NUREG/CP-OI38. US Nuclear Regulatory Commission, Washington, DC.

Netjasov, F. and Janic, M. (2008) A review of research on risk and safety modeling in civil aviation. *Journal of Air Transport Management*, **14**, 213–220.

Nguyen, H.T. (2006) *An Introduction to Random Sets*, CRC Press, Boca Raton, FL.

North, W. (2010) Probability theory and consistent reasoning. *Risk Analysis*, **30** (3), 377–380.

NRC (1975) Reactor Safety Study, an Assessment of Accident Risks. Wash 1400. Report NUREG-75/014. US Nuclear Regulatory Commission, Washington, DC.

O'Hagan, A. and Oakley, J.E. (2004) Probability is perfect, but we can't elicit it perfectly. *Reliability Engineering and System Safety*, **85**, 239–248.

Parry, G. and Drouin, M.T. (2009) Risk-Informed Regulatory Decision-Making at the U.S. NRC: Dealing with model uncertainty. US Nuclear Regulatory Commission, Washington, DC.

Paté-Cornell, M.E. (1996) Uncertainties in risk analysis: six levels of treatment. *Reliability Engineering and System Safety*, **54** (2–3), 95–111.

Rechard, R.P. (1999) Historical relationship between performance assessment for radioactive waste disposal and other types of risk assessment. *Risk Analysis*, **19** (5), 763–807.

Rechard, R.P. (2000) Historical background on performance assessment for the waste isolation pilot plant. *Reliability Engineering and System Safety*, **69** (3), 5–46.

Renn, O. (1998) Three decades of risk research: accomplishments and new challenges. *Journal of Risk Research*, **1** (1), 49–71.

Renn, O. (2005) Risk Governance. White Paper no. 1, International Risk Governance Council, Geneva.

Renn, O. (2008) *Risk Governance: Coping with Uncertainty in a Complex World*, Earthscan, London.

RESS (2004) Special issue on alternative representations of epistemic uncertainty. *Reliability Engineering and System Safety*, **88** (1–3) (eds. J.C. Helton and W.L. Oberkampf).

Saltelli, A., Ratto, M., Andres, T. *et al.* (2008) *Global Sensitivity Analysis: The Primer*, John Wiley & Sons, Inc., Hoboken, NJ.

Shafer, G. (1976) *A Mathematical Theory of Evidence*, Princeton University Press, Princeton, NJ.

Singpurwalla, N.D. (2006) *Reliability and Risk: A Bayesian Perspective*, John Wiley & Sons, Ltd, Chichester.

Smets, P. (1994) What is Dempster–Shafer's model?, in *Advances in The Dempster–Shafer Theory of Evidence* (eds. R.R. Yager, M. Fedrizzi, and J. Kacprzyk), John Wiley & Sons, New York, pp. 5–34.

Tickner, J. and Kriebel, D. (2006) The role of science and precaution in environmental and public health policy, in *Implementing the Precautionary Principle* (eds E. Fisher, J. Jones, and R.von Schomberg), Edward Elgar Publishing, Northampton, MA, USA.

US National Research Council (1996) *Understanding Risk: Informing Decisions in a Democratic Society*, National Academy Press (NAP), Washington, DC.

US National Research Council (of the National Academies) (2008) *Public Participation in Environmental Assessment and Decision Making*, National Academy Press (NAP), Washington, DC.

Vesely, W.E. and Apostolakis, G.E. (1999) Developments in risk-informed decision-making for nuclear power plants. *Reliability Engineering and System Safety*, **63**, 223–224.

Vinnem, J.E. (2007) *Offshore Risk Assessment: Principles, Modelling and Applications of QRA Studies*, 2nd edn, Springer, London.

Vinnem, J.E. (2010) Risk analysis and risk acceptance criteria in the planning processes of hazardous facilities – a case of an LNG plant in an urban area. *Reliability Engineering and System Safety*, **95** (6), 662–670.

Vose, D. (2008) *Risk Analysis: A Quantitative Guide*, 3rd edn, John Wiley & Sons, Ltd, Chichester.

Walley, P. (1991) *Statistical Reasoning with Imprecise Probabilities*, Chapman & Hall, New York.

Winkler, R.L. (1996) Uncertainty in probabilistic risk assessment. *Reliability Engineering and System Safety*, **85**, 127–132.

Zio, E. (2007) *An Introduction to the Basics of Reliability and Risk Analysis*, World Scientific, Hackensack, NJ.

Zio, E. (2009) Reliability engineering: old problems and new challenges. *Reliability Engineering and System Safety*, **94**, 125–141.

Zio, E. and Apostolakis, G.E. (1996) Two methods for the structured assessment of model uncertainty by experts in performance assessments of radioactive waste repositories. *Reliability Engineering and System Safety*, **54**, 225–241.

Part II

METHODS

In the second part of the book we review and discuss several methods for representing uncertainty, as well as associated approaches for the treatment of uncertainty in the context of risk assessment. A key point of reference is the following list of features specifying what is expected from a mathematical representation of uncertainty (Bedford and Cooke, 2001, p. 20):

- Axioms: Specifying the formal properties of uncertainty.

- Interpretations: Connecting the primitive terms in the axioms with observable phenomena.

- Measurement procedures: Providing, together with supplementary assumptions, practical methods for interpreting the axiom system.

We will particularly address the second bullet point. For the purpose of illustration, we will make use of some simple examples in the presentation. As the intention is to highlight key conceptual features and ideas, the examples in Part II are kept rather basic compared to the more comprehensive applications in Part III. The first example is introduced below.

Example II.1

We consider the operation of a chemical reactor (Figure II.1) in which different scenarios may occur after failure of the reactor cooling system.

Let D denote the failure event of the cooling system. The resulting scenarios are classified with increasing criticality according to the maximum temperature, T_{max}, reached within the reactor as a consequence of the failure event (Figure II.2):

- Safe (SA), characterized by a maximum temperature in the range 100–150 °C, that is, $T_{max} \in [100, 150)$°C

- Marginal (MA), if $T_{max} \in [150, 200)$°C

Uncertainty in Risk Assessment: The Representation and Treatment of Uncertainties by Probabilistic and Non-Probabilistic Methods, First Edition. Terje Aven, Piero Baraldi, Roger Flage and Enrico Zio.
© 2014 John Wiley & Sons, Ltd. Published 2014 by John Wiley & Sons, Ltd.

- Critical (CR), if $T_{max} \in [200, 300)°C$

- Catastrophic (CA), if $T_{max} \in [300, 500)°C$

Values of T_{max} lower than 100 °C and larger than 500 °C cannot be reached for physical reasons.

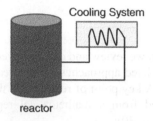

Figure II.1   Scheme of the chemical reactor and the cooling system.

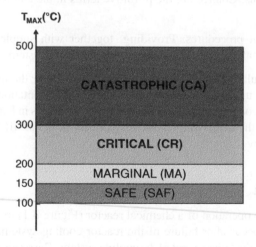

Figure II.2   Classification of the scenarios according to the maximum temperature T_{max} reached within the reactor as a consequence of the failure event.

We first consider the probabilistic representation, which is most commonly used in the risk assessment setting and upon which the other representations are built.

2

Probabilistic approaches for treating uncertainty

Probability has been, and continues to be, a fundamental concept in the context of risk assessment. Unlike the other measures of uncertainty considered in this book, probability is a single-valued measure.

The common axiomatic basis of probability is found in Kolmogorov's (1933) probability axioms. Let $A, A_1, A_2, \ldots$ denote events in a sample space S. The following probability axioms are assumed to hold:

$$
\begin{gathered}
0 \leq P(A) \\
P(S) = 1 \\
P(A_1 \cup A_2 \cup \ldots) = P(A_1) + P(A_2) + \ldots, \\
\text{if } A_i \cap A_j = \varnothing \quad \text{for all } i \text{ and } j, \ i \neq j.
\end{gathered}
\tag{2.1}
$$

These axioms specify that probability is a non-negative, normalized, and additive measure.

Different types of operational or conceptual interpretations of probability can be distinguished. Among the most common are:

- Classical

- Relative frequency

- Subjective

- Logical.

Uncertainty in Risk Assessment: The Representation and Treatment of Uncertainties by Probabilistic and Non-Probabilistic Methods, First Edition. Terje Aven, Piero Baraldi, Roger Flage and Enrico Zio.

Modern probability theory is not based on a particular interpretation of probability, although its standard language is best suited to the classical and relative frequency interpretations (Aven, 2012). In the context of risk assessment, the relative frequency and subjective interpretations are most common. Relative frequency probability is broadly recognized as the proper representation of variation in large populations (aleatory uncertainty), also among proponents of alternative representations of uncertainty; see, for example, Baudrit, Dubois, and Guyonnet (2006) and Ferson and Ginzburg (1996). The subjective interpretation, where probability is understood as expressing a degree of belief and measuring epistemic uncertainty, is also widely applied in risk assessment (Bedford and Cooke, 2001). The classical and logical interpretations of probability have less relevance in the context of risk assessment, as will be seen in the coming sections where we review and discuss the interpretations mentioned above. The review, which to a large extent is taken from or based on Aven (2013a), covers also the Bayesian subjective probability framework, which through the concept of chance provides a link between subjective probability and limiting relative frequencies.

2.1 Classical probabilities

The classical interpretation of probability dates back to de Laplace (1812). It applies only in situations involving a finite number of outcomes which are equally likely to occur. The probability of an event A is equal to the ratio between the number of outcomes resulting in A and the total number of possible outcomes, that is,

$$P(A) = \text{Number of outcomes resulting in } A/\text{Total number of possible outcomes.}$$

$$(2.2)$$

Consider as an example the tossing of a die. Then $P(\text{the die shows "1"}) = 1/6$, as there are six possible outcomes which are equally likely to appear and only one outcome which gives the event that the die shows "1." The requirement for each outcome to be equally likely is crucial for the understanding of this interpretation and has been subject to much discussion in the literature. A common perspective taken is that this requirement is met if there is no evidence favoring some outcomes over others. This is the so-called "principle of indifference," also sometimes referred to as the "principle of insufficient reason." In other words, classical probabilities are appropriate when the evidence, if there is any, is symmetrically balanced (Hajek, 2001), such as we may have when throwing a die or playing a card game.

The classical interpretation is, however, not applicable in most real-life situations beyond random gambling and sampling, as we seldom have a finite number of outcomes which are equally likely to occur. The discussion about the principle of indifference is interesting from a theoretical point of view, but a probability concept based solely on this principle is not so relevant in a context where we search for a concept of probability that can be used in a wide range of applications.

Example 2.1

Referring to Example II.1 introduced at the beginning of Part II; the classical interpretation would only be applicable if there were no evidence to suggest that any of the scenarios following a cooling system failure event is more likely than the rest. For the scenarios of different criticality considered in the example we would, however, expect different values of probabilities (smaller for the more critical ones) as the design and operation of reactors are typically based on several layers of barriers implemented to avoid unwanted initiating events developing into severely critical consequences (this is referred to as the multiple barrier implementation of the defense-in-depth philosophy (Zio, 2007)). Hence the classical interpretation is not applicable in this case.

2.2   Frequentist probabilities

The frequentist probability of an event A, denoted $P_f(A)$, is defined as the fraction of times event A occurs if the situation/experiment considered were repeated (hypothetically) an infinite number of times. Thus, if an experiment is performed n times and event A occurs n_A times out of these, then $P_f(A)$ is equal to the limit of n_A/n as n tends to infinity (tacitly assuming that the limit exists), that is,

$$P_f(A) = \lim_{n \to \infty} \frac{n_A}{n}. \qquad (2.3)$$

Taking a sample of repetitions of the experiment, event A occurs in some of the repetitions and not in the rest. This phenomenon is attributed to "randomness," and asymptotically the process generates a fraction of successes, the "true" probability $P_f(A)$, which describes quantitatively the aleatory uncertainty (i.e., variation) about the occurrence of event A. In practice, it is of course not possible to repeat a situation an infinite number of times. The probability $P_f(A)$ is a model concept used as an approximation to a real-world setting where the population of units or number of repetitions is always finite. The limiting fraction $P_f(A)$ is typically unknown and needs to be estimated by the fraction of occurrences of A in the finite sample considered, producing an estimate $P_f^*(A)$.

A frequentist probability is thus a mind-constructed quantity – a model concept founded on the law of large numbers which says that frequencies n_A/n converge to a limit under certain conditions: that is, the probability of event A exists and is the same in all experiments, and the experiments are independent. These conditions themselves appeal to probability, generating a problem of circularity. One approach to deal with this problem is to assign the concept of probability to an individual event by embedding the event into an infinite class of "similar" events having certain "randomness" properties (Bedford and Cooke, 2001, p. 23). This leads to a somewhat complicated framework for understanding the concept of probability; see the discussion in van Lambalgen (1990). Another approach, and a common way of looking at

Figure 2.1   Repetition of the experiment of tossing a coin.

probability, is to simply assume the existence of the probability $P_f(A)$, and then refer to the law of large numbers to give $P_f(A)$ the limiting relative frequency interpretation. Starting from Kolmogorov's axioms (as in the second paragraph of the introduction to Part II; see also Bedford and Cooke 2001, p. 40) and the concept of conditional probability, as well as presuming the existence of probability, the theory of probability is derived, wherein the law of large numbers constitutes a key theorem which provides the interpretation of the probability concept.

The so-called propensity interpretation is yet another way of justifying the existence of a probability (SEP, 2009). This interpretation specifies that probability should primarily be thought of as a physical characteristic: The probability is a propensity of a repeatable experimental set-up which produces outcomes with limiting relative frequency $P_f(A)$. Consider as an example a coin. The physical characteristics of the coin (weight, center of mass, etc.) are such that when tossing the coin over and over again the fraction of heads will be p (Figure 2.1). The literature on probability shows that the existence of a propensity is controversial (Aven, 2013a). However, from a conceptual point of view the idea of a propensity may be no more difficult to grasp than the idea of an infinite number of repetitions of experiments. If the framework of frequentist probability is accepted, that is, if referring to an infinite sequence of similar situations makes sense, the propensity concept should also be accepted as it essentially expresses that such a framework exists.

Thus, for gambling situations and when dealing with fractions in large populations of similar items, the frequentist probability concept makes sense, as a model concept. Of course, if a die is thrown over and over again a very large number of times its physical properties will eventually change, so the idea of "similar experiments" is questionable. Nevertheless, for all practical purposes we can carry out (in theory) a large number of trials, say 100 000, without any physical changes in the experimental set-up, and that is what is required for the concept to be meaningful. The same is true if we consider a population of (say) 100 000 human beings belonging to a specific category, say men in the age range 20 to 30 years old in a specific country.

It is not surprising, then, that frequency probabilities are so commonly and widely adopted in practice. The theoretical concept of a frequentist probability is introduced, often in the form of a probability model – for example, the binomial, normal (Gaussian), or Poisson distributions – and statistical analysis is carried out to estimate the frequentist probabilities (more generally the parameters of the probability models) and to study the properties of the estimators using well-established statistical theory. However, the types of situations that are captured by this framework are limited. As noted by Singpurwalla (2006, p. 17), the concept of frequency probabilities "is applicable to only those situations for which we can conceive of a repeatable experiment." This excludes many situations and events. Consider for example events such as the rise of sea level over the next 20 years, the guilt or innocence of an accused individual, or the occurrence or not of a disease in a specific person with a specific history.

What does it mean that the situations under consideration are "similar?" The "experimental conditions" cannot be identical, since we would then observe exactly the same outcome and the ratio n_A/n would be either 1 or 0. What kind of variation between the experiments is allowed? This is often difficult to specify and makes it challenging to extend frequentist probabilities to include real-life situations. Consider as an example the frequentist probability that a person V contracts a specific disease D. What should be the population of similar persons in this situation? If we include all men/women of his/her age group we get a large population, but many of the people in this population may not be very "similar" to person V. We may reduce the population to increase the similarity, but not too much as that would make the population very small and hence inadequate for defining the frequentist probability. This type of dilemma is faced in many types of modeling and a balance has to be made between different concerns: similarity (and hence relevance) vs. population size (and hence validity of the frequentist probability concept, as well as data availability).

Example 2.1 (continued)

For the chemical reactor example, the relative frequency interpretation requires the introduction of an infinite population of "similar" years of operation of the reactor in question or, alternatively, an infinite population of similar reactors (under similar operating conditions). The following model concepts (parameters) are introduced: the fraction of years when the cooling system fails is denoted $P_f(D)$, and the fraction of years where scenarios of the different criticality classes occur are denoted $P_f(SA)$, $P_f(MA)$, $P_f(CR)$, and $P_f(CA)$. These quantities are unknown, and the standard statistical way to proceed is to establish estimates and describe the associated uncertainty using confidence intervals. A $p \times 100\%$ confidence interval is defined by

$$P_f(Y_1 \leq \theta \leq Y_2) = p, \tag{2.4}$$

where θ is a parameter and Y_1 and Y_2 random variables. There might be data available to provide an estimate $P_f^*(D)$ of the probability of a cooling system

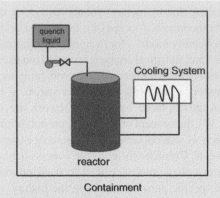

Figure 2.2   Reactor system and barriers.

failure event D; however, for the rarer scenarios, such as the catastrophic one (CA), data directly on the overall event may not be available. A model g then needs to be introduced which links the conditional frequentist probability of a catastrophic CA scenario (given a cooling system failure event D), $P_f(CA \mid D)$, with frequentist probabilities of barrier failures. An event tree may be used for this purpose. Consider for example the following two barriers, illustrated in Figure 2.2:

1. Protection system: A quench system which stops the reaction by pumping a quench liquid inside the reactor.

2. Mitigation system: A containment which in case of failure of the quench system should mitigate the external damage.

Upon a cooling system failure event D, the quench system is the first barrier to avoid overheating of the reactor. If the quench system fails, a containment system is intended to mitigate external damage. If both these barriers fail, a catastrophic scenario will occur. The logic of this is illustrated in the event tree of Figure 2.3, where B_1 denotes failure of the quench system and B_2 failure of the containment system.

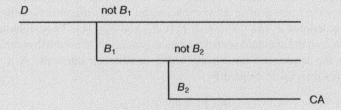

Figure 2.3   Event tree of the reactor system barriers.

Let $q = (q_1, q_2)$ be the vector of barrier failure probabilities, where $q_1 = P_f(B_1 \mid D)$ denotes the frequentist probability of quench system failure given a cooling system failure, and $q_2 = P_f(B_2 \mid D, B_1)$ the frequentist probability of containment system failure given a cooling system failure and quench system failure. From the event tree in Figure 2.3. we construct a model g of $P_f(CA \mid D)$ by $g(q) = q_1 q_2$. The barrier failure probabilities can be estimated from test data, and an estimate of the probability of a catastrophic failure is obtained as

$$P_f^*(CA) = P_f^*(D) P_f^*(CA \mid D) = P_f^*(D) g(q^*).$$

Based on the data used to estimate the frequentist probability $P_f(D)$, an associated confidence interval for this probability can be given. On the other hand, calculating a confidence interval for the unconditional frequentist probability $P_f(CA)$ is not so straightforward since data directly on the CA scenario is not available.

2.3 Subjective probabilities

The theory of subjective probability was proposed independently and at about the same time by Bruno de Finetti in *Fondamenti Logici del Ragionamento Probabilistico* (de Finetti, 1930) and Frank Ramsey in *The Foundations of Mathematics* (Ramsey, 1931); see Gillies (2000).

A subjective probability – sometimes also referred to as a judgmental or knowledge-based probability – is a purely epistemic description of uncertainty as seen by the assigner, based on his or her background knowledge. In this view, the probability of an event A represents the degree of belief of the assigner with regard to the occurrence of A. Hence, a probability assignment is a numerical encoding of the state of knowledge of the assessor, rather than a property of the "real world."

It is important to appreciate that, irrespective of interpretation, any subjective probability is considered to be conditional on the background knowledge K that the assignment is based on. They are "probabilities in the light of current knowledge" (Lindley, 2006). This can be written as $P(A \mid K)$, although the writing of K is normally omitted as the background knowledge is usually unmodified throughout the calculations. Thus, if the background knowledge changes, the probability might also change. Bayes' theorem (see Section 2.4) is the appropriate formal tool for incorporating additional knowledge into a subjective probability. In the context of risk assessment, the background knowledge typically and mainly includes data, models, expert statements, assumptions, and phenomenological understanding.

There are two common interpretations of a subjective probability, one making reference to betting and another to a standard for measurement. The betting interpretation and related interpretations dominate the literature on subjective probabilities, especially within the fields of economy and decision analysis, whereas the standard for measurement is more common among risk and safety analysts.

2.3.1   Betting interpretation

If derived and interpreted with reference to betting, the probability of an event A, denoted $P(A)$, equals the amount of money that the person assigning the probability would be willing to bet if a single unit of payment were given in return in case event A were to occur, and nothing otherwise. The opposite must also hold: that is, the assessor must be willing to bet the amount $1-P(A)$ if a single unit of payment were given in return in case A were not to occur, and nothing otherwise. In other words, the probability of an event is the price at which the person assigning the probability is neutral between buying and selling a ticket that is worth one unit of payment if the event occurs and worthless if not (Singpurwalla, 2006). The two-sidedness of the bet is important in order to avoid a so-called Dutch book, that is, a combination of bets (probabilities) that the assigner would be committed to accept but which would lead him or her to a sure loss (Dubucs, 1993). A Dutch book can only be avoided by making so-called coherent bets, meaning bets that can be shown to obey the set of rules that probabilities must obey. In fact, the rules of probability theory can be derived by taking the avoidance of Dutch books as a starting point (Lindley, 2000).

Consider the event A, defined as the occurrence of a specific type of nuclear accident. Suppose that a person specifies the subjective probability $P(A)=0.005$. Then, according to the betting interpretation, this person is expressing that he/she is indifferent between:

- receiving (paying) €0.005; and

- taking a gamble in which he/she receives (pays) €1 if A occurs and €0 if A does not occur.

If the unit of money is €1000, the interpretation would be that the person is indifferent between:

- receiving (paying) €5; and

- taking a gamble where he/she receives (pays) €1000 if A occurs and €0 if A does not occur.

In practice, the probability assignment would be carried out according to an iterative procedure in which different gambles are compared until indifference is reached. However, as noted by Lindley (2006) (see also Cooke, 1986), receiving the payment in the nuclear accident example would be trivial if the accident were to occur (the assessor might not be alive to receive it). The problem is that there is a link between the probability assignment and value judgments about money (the price of the gamble) and the situation (the consequences of the accident). This value judgment has nothing to do with the uncertainties per se, or the degree of belief in the occurrence of event A.

2.3.2   Reference to a standard for uncertainty

A subjective probability can also be understood in relation to a standard for uncertainty, for example, making random withdrawals from an urn. If a person

Example 2.1 (continued)

A direct subjective probability assignment on the catastrophic (CA) scenario could be established without introducing frequency probabilities. The problem may be decomposed along the same lines as in the case of frequency probabilities, based on the event tree model in Figure 2.3, by introducing the model $I(CA) = g(D, B) = I(D, B_1, B_2)$ linking the occurrence of a cooling system failure (D) with the occurrence of a catastrophic scenario (CA) through the barrier failure events $B = (B_1, B_2)$. Here I is the indicator function equal to 1 if the argument is true and 0 otherwise. Then, we obtain

$$P(CA) = E[I(D, B_1, B_2)] = P(D, B_1, B_2) = P(D)P(B_1|D)P(B_2|D, B_1),$$

where E is the expected value operator.

assigns a probability of 0.1 (say) to an event A, then this person compares his/her uncertainty (degree of belief) in the occurrence of A to drawing a specific ball from an urn containing 10 balls. The uncertainty (degree of belief) is the same. More generally, the probability $P(A)$ is the number such that the uncertainty about the occurrence of A is considered equivalent, by the person assigning the probability, to the uncertainty about the occurrence of some standard event, for example, drawing, at random, a red ball from an urn that contains $P(A) \times 100\%$ red balls (see, e.g., Lindley, 2000, 2006; Bernardo and Smith, 1994).

As for the betting interpretation, the interpretation with reference to an uncertainty standard can be used to deduce the rules of probability; see Lindley (2000). These rules are typically referred to as axioms in textbooks on probability, but they are not axioms here, since they are deduced from more basic assumptions linked to the uncertainty standard; see Lindley (2000, p. 299). Whether the probability rules are deduced or taken as axioms may not be so important to applied probabilists. The main point is that these rules apply, and the uncertainty standard provides an easily understandable way of defining and interpreting subjective probability where uncertainty/probability and utility/value are separated. The rules of probability reduce to the rules governing proportions, which are easy to communicate.

2.4 The Bayesian subjective probability framework

The so-called Bayesian framework has subjective probability as a basic component, and the term "probability" is reserved for, and always understood as, a degree of belief. Within this framework, the term "chance" is used by some authors (e.g., Lindley, 2006; Singpurwalla, 2006; Singpurwalla and Wilson, 2008) for the limit of a relative frequency in an exchangeable, infinite Bernoulli series. A chance distribution is the limit of an empirical distribution function (Lindley, 2006). Two random quantities Y_1 and Y_2 are said to be exchangeable if for all values y_1 and y_2 that

Y_1 and Y_2 may take, we have

$$P(Y_1 \leq y_1, Y_2 \leq y_2) = P(Y_1 \leq y_2, Y_2 \leq y_1). \tag{2.5}$$

That is, the probabilities remain unchanged (invariant) when switching (permuting) the indices. The relationship between subjective probability and the chance concept is given by the so-called representation theorem of de Finetti; see, for example, Bernardo and Smith (1994, p. 172). Roughly speaking, this theorem states that if an exchangeable Bernoulli series can be introduced, one may act as though frequency probabilities exist.

In the case when a chance p of an event A may be introduced, we have $P(A|p)=p$. That is, the probability of an event A for which the associated chance is known is simply equal to the value of the chance. For unknown p, observations of outcomes in similar (exchangeable) situations would not be considered independent of the situation of interest, since the observations would provide more information about the value of p. On the other hand, for known p the outcomes would be judged as independent, since nothing more could be learned about p from additional observations. Hence, conditional on p, the outcomes are independent, but unconditionally they are not – they are exchangeable. In practice, the value of a chance p is in most cases unknown and the assessor expresses his or her uncertainty about the value of p by a probability distribution $H(p') = P(p \leq p')$. As in the example of Section 1.1.3, the probability of A can then be derived as

$$P(A) = \int_0^1 P(A|p') \, dH(p') = \int_0^1 p' \, dH(p'). \tag{2.6}$$

The probability $P(A)$ in this equation characterizes uncertainty about the occurrence of event A, given the background knowledge K (suppressed from the notation), which includes a judgment of exchangeability which allows for the equivalence relation $P(A|p)=p$, as well as the information contained in $H(p')$. Consequently, the uncertainty about p does not make $P(A)$ uncertain.

More generally, if a chance distribution is known, then under judgment of exchangeability the probability distribution is equal to the chance distribution, which is analogous to setting $P(A|p)=p$. If the chance distribution is unknown, a "prior" probability distribution is established over the chance distribution (parameters), updated to a "posterior" distribution upon receiving new information, and the "predictive" probability distribution may be established by using the law of total probability, as illustrated in (2.6). The updating is performed using Bayes' rule, which states that

$$P(A|B) = \frac{P(B|A)P(A)}{P(B)} \tag{2.7}$$

for $P(B) > 0$.

Example 2.4

The frequentist probabilities introduced in Section 2.2 would be introduced in a Bayesian setting as chances. Let u denote the chance of a catastrophic (CA) scenario; also let v_1 denote the chance of a cooling system failure (chance of event D), v_2 the chance of a quenching system failure (chance of event B_1), and v_3 the chance of a containment system failure (chance of event B_2). Then we have a chance model g as follows:

$$u = g(v) = v_1 v_2 v_3,$$

where $v = (v_1, v_2, v_3)$.

The values of these chances are uncertain and the uncertainty is described by a subjective probability distribution $H(v') = P(v \leq v')$. Analogously to (2.6), the probability of the catastrophic scenario is given as

$$P(CA) = E[u] = \int g(v) \mathrm{d}H(v) = \int v_1 v_2 v_3 \, \mathrm{d}H(v).$$

In addition a probability distribution of u may be derived, describing the degree of uncertainty on the value of the chance of a CA scenario. From this probability distribution a credibility interval may then be derived, say a 90% credibility interval $[a, b]$ specifying that with a 90% probability the actual value of u is in the interval $[a, b]$, where a and b are fixed quantities.

2.5   Logical probabilities

Logical probabilities were first proposed by Keynes (1921) and later taken up by Carnap (1922, 1929). The idea is that probability expresses an objective logical relation between propositions – a sort of "partial entailment." There is a number in the interval $[0,1]$, denoted $P(H \mid E)$, which measures the objective degree of logical support that evidence E gives to hypothesis H (Franklin, 2001). As stated by Franklin (2001), this view on probability has an intuitive initial attractiveness in representing a level of agreement found when scientists, juries, actuaries, and so on evaluate hypotheses in the light of evidence. However, as described by Aven (2013a), the notion of partial entailment has never received a satisfactory interpretation, and also both Cowell *et al.* (1999) and Cooke (2004) conclude that this interpretation of probability cannot be justified.

3

Imprecise probabilities for treating uncertainty

As described in Chapter 1, it has been argued that in situations of poor knowledge, representations of uncertainty based on lower and upper probabilities are more appropriate than precise probabilities. This chapter presents imprecise probabilityes for treating uncertainty, largely taken from Aven (2011c).

The first theoretical foundation for imprecise probability was laid down by Boole (1854). More recently, Peter M. Williams developed a mathematical foundation for imprecise probabilities based on de Finetti's betting interpretation of probability (de Finetti, 1974). This foundation was further developed independently by Kuznetsov (1991) and Walley (1991).

The term "imprecise probability" brings together a variety of different theories (Coolen, Troffaes, and Augustin, 2010). It is used to refer to a generalization of probability theory based on the representation of the uncertainty about an event A through the use of a lower probability $\underline{P}(A)$ and an upper probability $\overline{P}(A)$, where $0 \leq \overline{P}(A) \leq \overline{P}(A) \leq 1$. The imprecision in the representation of event A is defined as (Coolen, 2004)

$$\Delta P(A) = \overline{P}(A) - \underline{P}(A). \tag{3.1}$$

The special case with $\underline{P}(A) = \overline{P}(A)$ (and hence $\Delta P(A) = 0$) for all events A leads to conventional probability theory, whereas the case with $\underline{P}(A) = 0$ and $\overline{P}(A) = 1$ (and hence $\Delta P(A) = 1$) represents complete lack of knowledge, with a flexible continuum in between (Coolen, Troffaes, and Augustin, 2010).

Uncertainty in Risk Assessment: The Representation and Treatment of Uncertainties by Probabilistic and Non-Probabilistic Methods, First Edition. Terje Aven, Piero Baraldi, Roger Flage and Enrico Zio.
© 2014 John Wiley & Sons, Ltd. Published 2014 by John Wiley & Sons, Ltd.

Example 3.1

The uncertainty related to the occurrence of a catastrophic (CA) scenario given a failure event D of the cooling system could be described by assigning, say, $\underline{P}(CA \mid D) = 0.01$ and $\overline{P}(CA \mid D) = 0.1$. The associated imprecision would then be $\Delta P(CA \mid D) = 0.1 - 0.01 = 0.09$.

A range of generalizations of classical probability theory have been presented, in terms of axiom systems and further concepts and theorems (Coolen, 2004). Many researchers agree that the most complete framework of lower and upper probabilities is offered by Walley's (1991) theory of imprecise probability and the closely related theory of interval probability by Weichselberger (2000), where the former emphasizes developments via coherent subjective betting behavior (see the paragraph below) and the latter is formulated more as a generalization of Kolmogorov's axioms of classical probability (Coolen, 2004). Axioms and further concepts are far more complex for imprecise probability than for classical precise probability; for example, the concept of conditional probability does not have a unique generalization to imprecise probability (Coolen, 2004).

Following de Finetti's betting interpretation, Walley (1991) has proposed to interpret lower probability as the maximum price for which one would be willing to buy a bet which pays 1 if A occurs and 0 if not, and the upper probability as the minimum price for which one would be willing to sell the same bet.

Alternatively, lower and upper probabilities may be interpreted with reference to an uncertainty standard, as introduced in Section 2.3. Such an interpretation is indicated by Lindley (2006, p. 36). Consider the assignment of a subjective probability $P(A)$ and suppose that the assigner states that his or her degree of belief is reflected by a probability greater than a comparable urn chance of 0.10 and less than an urn chance of 0.5. The analyst is not willing to make a more precise assignment. Then the interval $[0.1, 0.5]$ can be considered an imprecision interval for the probability $P(A)$.

Of course, even if the assessor assigns a single probability $P(A) = 0.3$, this can be understood as an imprecise probability interval equal to say $[0.26, 0.34]$ (since a number in this interval is equal to 0.3 when only one digit is displayed), interpreted analogously as the $[0.2, 0.5]$ interval above. Imprecision is always an issue in a practical context. This type of imprecision is typically seen to be a result of measurement problems (cf. the third bullet point in the list presented at the beginning of Part II). Lindley (2006) argues that the use of lower and upper probabilities confuses the concept of interpretation with the measurement procedures (in his terminology, the "concept" of measurement with the "practice" of measurement). The reference to the urn standard provides a norm, and measurement problems may make the assessor unable to behave according to it.

Other researchers and analysts are more positive with respect to the need for using imprecise/interval probabilities in practice; see the discussions in, for example,

Dubois (2010), Aven and Zio (2011), Walley (1991), and Ferson and Ginzburg (1996). Imprecision intervals are required to reflect phenomena involved in uncertainty assessments, for example, experts that are not willing to represent their knowledge and uncertainty more precisely than by using probability intervals.

Imprecise probabilities can also be linked to the relative frequency interpretation (Coolen and Utkin, 2007). In the simplest case the "true" frequentist probability $p = P_f(A)$ is in the specified interval with certainty (i.e., with subjective probability 1). More generally, and in line with a subjective interpretation of imprecision intervals, a two-level uncertainty characterization can be formulated (see, e.g., Kozine and Utkin, 2002). The interval $[\underline{P}(A), \overline{P}(A)]$ is an imprecision interval for the subjective probability $P(A) = P(a \leq p \leq b)$, where a and b are fixed quantities. In the special case that $\underline{P}(A) = \overline{P}(A) = q$ (say), we have a $q \times 100\%$ credibility interval (cf. Section 2.4) for p specifying that with a subjective probability q the true value of p is in the interval $[a, b]$; cf. Example 2.4 in Chapter 2.4.

4

Possibility theory for treating uncertainty

Possibility theory deals with uncertainty characterization in the case of incomplete information. It differs from probability theory (which uses a single probability measure) in that it uses a pair of dual set functions called possibility and necessity measures. The background and scope of possibility theory is described as follows by Dubois and Prade (2007):

> The name Theory of Possibility was coined by (Zadeh, 1978), inspired by (Gaines and Kohout, 1975). In Zadeh's view, possibility distributions were meant to provide a graded semantics to natural language statements. However, possibility and necessity measures can also be the basis of a full-fledged representation of partial belief that parallels probability (Dubois and Prade, 1988). Then, it can be seen either as a coarse, non-numerical version of probability theory, or as a framework for reasoning with extreme probabilities, or yet as a simple approach to reasoning with imprecise probabilities (Dubois, Nguyen, and Prade, 2000).

In the following we first review some basics of possibility theory, largely taken from or based on Aven (2011c), and then present some approaches to constructing possibility distributions.

4.1 Basics of possibility theory

A central component in possibility theory is the possibility function π. For each y in a set S, $\pi(y)$ expresses the degree of possibility of y. When $\pi(y) = 0$ for some y, it means that the outcome y is considered an impossible situation, whereas when $\pi(y) = 1$ for some y, it means that the outcome y is possible; that is, is just unsurprising, normal,

Uncertainty in Risk Assessment: The Representation and Treatment of Uncertainties by Probabilistic and Non-Probabilistic Methods, First Edition. Terje Aven, Piero Baraldi, Roger Flage and Enrico Zio.
© 2014 John Wiley & Sons, Ltd. Published 2014 by John Wiley & Sons, Ltd.

usual (Dubois, 2006). This is a much weaker statement than a probability equal to 1. Since one of the elements of S is the true value, it is assumed that $\pi(y) = 1$ for at least one y.

To interpret a possibility function π, we consider the pair of necessity/possibility measures $[N, \Pi]$ that it induces. The possibility of an event A, $\Pi(A)$, is defined by

$$\Pi(A) = \sup_{y \in A} \pi(y), \tag{4.1}$$

and the associated necessity measure, $N(A)$, is defined by

$$N(A) = 1 - \Pi(\bar{A}) = \inf_{y \notin A}(1 - \pi(y)), \tag{4.2}$$

where $\bar{A}$ is the complement of A. The possibility measure satisfies the properties

$$\Pi(\varnothing) = 0 \tag{4.3}$$

$$\Pi(S) = 1 \tag{4.4}$$

$$\Pi(A \cup B) = \max\{\Pi(A), \Pi(B)\}, \tag{4.5}$$

for disjoint sets A and B, where $\varnothing$ is the empty set. From these properties and (4.2), properties of the necessity measure can be derived, for example,

$$N(A \cap B) = \min\{N(A), N(B)\}. \tag{4.6}$$

Let $\mathbf{P}(\pi)$ be a family of probability distributions such that for all events A,

$$N(A) \le P(A) \le \Pi(A). \tag{4.7}$$

Then

$$N(A) = \inf_{\mathbf{P}(\pi)} P(A) \quad \text{and} \quad \Pi(A) = \sup_{\mathbf{P}(\pi)} P(A). \tag{4.8}$$

Thus the possibility and necessity measures of an event can be interpreted as upper and lower limits, respectively, for the probability of the same event. Using subjective probabilities with reference to a standard for uncertainty, the bounds reflect that the analyst is not willing to precisely assign a single-valued probability.

A typical example of the implementation of a possibilistic representation is the following (Anoop and Rao, 2008; Baraldi and Zio, 2008). Consider an uncertain parameter Θ. Based on the definition of the parameter, we know that it can take values in the range $[1, 3]$. Furthermore, the most likely value of the parameter is 2. To represent this information a triangular possibility distribution on the interval $[1, 3]$ is used, with a peak value at $\theta = 2$, as illustrated in Figure 4.1.

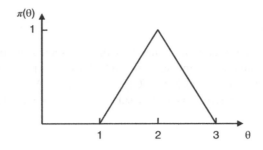

Figure 4.1   Triangular possibility distribution for a parameter x on the interval [1, 3], with maximum value at θ = 2.

From (4.6), we can derive the cumulative necessity and possibility measures, $N(-\infty, \theta]$ and $\Pi(-\infty, \theta]$ associated with the possibility distribution in Figure 4.1, as shown in Figure 4.2. These measures are interpreted as the lower and upper cumulative probability distributions for the uncertain parameter Θ.

The triangular possibility distribution generates the bounds in Figure 4.2. Other possibility functions could also be specified on the basis of the present information; see Baudrit and Dubois (2006). However, as will be described in Section 3.2.2, the family of probability distributions induced by a triangular possibility distribution with range $[a, b]$ and core c contains all possible probability distributions with support $[a, b]$ and mode c. Starting by defining the bounds by, for example, the limits in Figure 4.2, interpretations can been given as for the interval probabilities discussed in Chapter 3. For example, the intervals are those obtained by the analyst as he or she is not able or willing to precisely assign his or her probability.

Possibility theory is suitable as a representation of uncertainty in situations where the available knowledge concerns nested subsets (Klir, 1998). Two sets are nested if one of the sets is a subset of the other. Sets in a sequence are nested if each subsequent set is contained in the next.

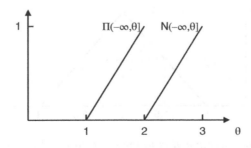

Figure 4.2   Bounds for the probability measures based on the possibility function in Figure 4.1.

Example 4.1

The sets of temperatures associated with the scenarios of class "SA" ($[100, 150)$ °C) and the joint scenario sets "SA or MA" ($[100, 200)$ °C), "SA or MA or CR" ($[100, 300)$ °C), and "SA or MA or CR or CA" ($[100, 500)$ °C) are nested, since each subsequent set is included in the next.

A unimodal possibility distribution, such as the one in Figure 4.2, can be seen as a nested set of intervals

$$A_\alpha = \left[\underline{y}_\alpha, \bar{y}_\alpha \right] = \{ y : \pi(y) \geq \alpha \}$$

of $\pi(y)$ which are usually referred to as α-cuts.

According to (4.1) and (4.2), the possibility and necessity measures associated with the α-cut A_α are

$$\Pi(A_\alpha) = 1, \tag{4.9}$$
$$N(A_\alpha) = 1 - \alpha. \tag{4.10}$$

Hence, the probability that the outcome of an uncertain quantity Y belongs to A_α is

$$P(Y \in A_\alpha) \geq 1 - \alpha. \tag{4.11}$$

From the possibility distribution in Figure 4.1 we have, for example, $A_{0.5} = [1.5, 2.5]$ (Figure 4.3) and we can conclude that $0.5 \leq P(1.5 \leq Y \leq 2.5) \leq 1$.

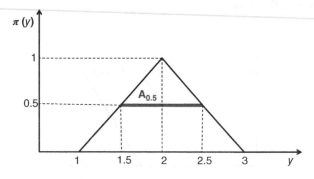

Figure 4.3   Interval corresponding to the α-cut $A_{0.5} = [1.5, 2.5]$ for the triangular possibility distribution of Figure 4.1.

4.2 Approaches for constructing possibility distributions

In this section, some techniques for constructing possibility distributions for unknown quantities subject to epistemic uncertainty are briefly described. The presentation is in part based on Baraldi *et al.* (2011). Section 4.2.1 presents a technique to build possibility distributions from one-sided probability intervals; Section 4.2.2 presents a justification for using triangular possibility distributions in cases where the range of possible values of the unknown quantity is known and a "preferred" value is specified; and finally, in Section 4.2.3, Chebyshev's inequality is used to build a possibility distribution when the available knowledge comprises the specification of the mean and standard deviation of an unknown quantity.

4.2.1 Building possibility distributions from nested probability intervals

Consider an unknown quantity Y and suppose that probabilities in the format of (4.11) are specified for intervals A_{α_i}, $i = 1, 2, \ldots, n$. Suppose also that the intervals are arranged in decreasing order (so that $\alpha_1 > \alpha_2 > \cdots > \alpha_n$) and the intervals are nested (so that $A_{\alpha_1} \subset A_{\alpha_2} \subset \cdots \subset A_{\alpha_n}$). According to (4.1) and (4.2) the interval A_{α_i} has a degree of necessity $N(A_{\alpha_i}) = 1 - \alpha_i$ and a degree of possibility $\Pi(A_{\alpha_i}) = 1$. The least restrictive possibility distribution consistent with the assigned probability intervals is defined by

$$\pi(y) = \begin{cases} 1 & \text{if } Y \in A_{\alpha_1} \\ \min_{i:y \notin A_{\alpha_i}} \alpha_i & \text{otherwise.} \end{cases} \qquad (4.12)$$

Equation (4.12) can be derived from (4.2) and says that for a value y contained in the largest interval A_{α_n}, one should identify the intervals A_{α_i} which do not contain y, select the one with the associated minimum value of α_i (i.e., the largest interval), and associate to $\pi(y)$ the value of α_i. Figure 4.4 shows the application of (4.12) to a case in which the intervals

$$A_{\alpha_1} = A_{0.75} = [1.75, 2.25], \quad A_{\alpha_2} = A_{0.5} = [1.5, 2.5], \quad A_{\alpha_3} = A_{0.25} = [1.25, 2.75],$$

$$A_{\alpha_4} = A_0 = [1, 3]$$

are provided.

In the example below, possibility theory is applied in a situation where the available information is slightly more refined than that considered in Section 4.3.

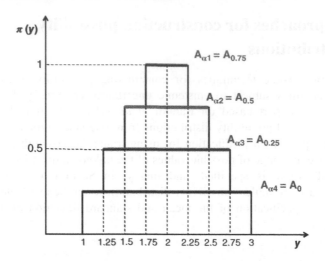

Figure 4.4 Possibility distributions obtained from the intervals $A_{\alpha_1} = A_{0.75} = [1.75, 2.25]$, $A_{\alpha_2} = A_{0.5} = [1.5, 2.5]$, $A_{\alpha_3} = A_{0.25} = [1.25, 2.75]$, *and* $A_{\alpha_4} = A_0 = [1, 3]$.

Example 4.1 (continued)

We consider the unknown quantity $T_{\max}$ representing the maximum temperature reached within a reactor as a consequence of reactor cooling system failure. Suppose that the available information for the scenarios which may occur after a cooling system failure is that the probability of a catastrophic (CA) scenario is greater than 0.2, and that the probability of a catastrophic (CA) or critical (CR) scenario is greater than 0.5. That is, it has been specified that $P(CA) > 0.2$ and $P(CA \text{ or } CR) = P(CA) + P(CR) > 0.5$. These assignments can be represented in a possibilistic format: define $A_1 = $ "CA", $A_2 = $ "CA or CR", and $A_3 = $ "CA or CR or MA or SA". According to the available probability assignments we have $N(A_1) = 0.2$, $N(A_2) = 0.5$, and $N(A_3) = 1$ (the latter equality holds since the failure event implies the occurrence of one of the four defined scenarios), and also $\Pi(A_1) = \Pi(A_2) = \Pi(A_3) = 1$ since no upper bounds on the probabilities have been provided. Considering the events A_1, A_2, and A_3 as having probabilities greater than or equal to $1 - \alpha_1 = 0.2$, and $1 - \alpha_3 = 1$, respectively, we obtain from (4.12) the possibility distribution of $T_{\max}$ shown in Figure 4.5 and the associated limiting cumulative distributions shown in Figure 4.6.

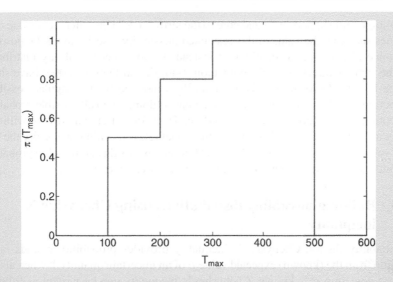

Figure 4.5   Possibility distribution associated with the unknown quantity T_{max}.

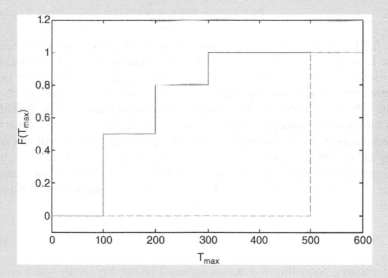

Figure 4.6   Limiting cumulative distributions associated with the possibility distribution in Figure 4.4. Solid and dashed lines represent upper and lower limiting cumulative distributions, respectively.

4.2.2   Justification for using the triangular possibility distribution

Consider a situation where the range $[a, b]$ of possible values for an unknown quantity Y is known, and where a "most likely" or "preferred" value c is also specified. This information seems to lead naturally to a representation by a triangular possibility distribution with support $[a, b]$ and core c (the term "core" here refers to a value or set of values such that the value of the associated possibility distribution is 1); however, the choice of a triangular distribution instead of any other possibility distribution with the same parameters needs motivation. Baudrit and Dubois (2006) have shown that the family of probability distributions $\mathbf{P}(\pi)$ induced by a triangular possibility distribution π with range $[a, b]$ and core c contains (dominates) all possible probability distributions with support $[a, b]$ and mode c. Therefore, if an expert is unwilling to specify a single probability distribution, but nonetheless is willing to specify the most likely value, the triangular possibility distribution ensures that all possible probability distributions consistent with the information given is covered.

4.2.3   Building possibility distributions using Chebyshev's inequality

In probability theory, Chebyshev's inequality provides probability bounds on the deviation from the (known) expected value μ of an uncertain quantity Y when also the standard deviation σ is known. Chebyshev's inequality can be written as follows:

$$P(|Y - \mu| \geq k\sigma) \leq 1/k^2, \tag{4.13}$$

for $k > 0$. The inequality is commonly used for proving convergence properties, but can also be used for representing uncertainty. For example, Chebyshev's inequality can be used to construct possibility distributions (Baudrit and Dubois, 2006). In fact, the use of continuous possibility distributions for representing probability families relies heavily on probabilistic inequalities. Such inequalities provide probability bounds for intervals forming a continuous nested family around a typical value. This nestedness property leads to an interpretation of the corresponding probability family as being induced by a possibility measure. Chebyshev's inequality defines a possibility distribution which dominates any probability density with given mean and variance. This allows the definition of a possibility distribution π by considering intervals $[\mu - k\sigma, \mu + k\sigma]$ as α-cuts of π and letting

$$\pi(\mu - k\sigma) = \pi(\mu + k\sigma) = 1/k^2. \tag{4.14}$$

The resulting possibility distribution defines a probability family which contains all probability distributions with mean μ and standard deviation σ, whether the probability distribution function is symmetric or not, or unimodal or not (Baudrit and Dubois, 2006).

5

Evidence theory for treating uncertainty

Evidence theory, also known as Dempster–Shafer theory or as the theory of belief functions, was established by Shafer (1976) for representing and reasoning with uncertain, imprecise, and incomplete information (Smets, 1994). It is a generalization of the Bayesian theory of subjective probability in the sense that it does not require probabilities for each event of interest, but bases belief in the truth of an event on the probabilities of other propositions or events related to it (Shafer, 1976). Evidence theory provides an alternative to the traditional manner in which probability theory is used to represent uncertainty by means of the specification of two degrees of likelihood, belief and plausibility, for each event under consideration. Evidence theory is introduced here with reference to Example 5.1, assuming that we have available knowledge in a different format than that considered in Chapters 3 and 4. The presentation in this chapter is in part taken from or based on Flage *et al.* (2009).

Example 5.1

With respect to the occurrence of a scenario of critical or catastrophic class, we assume we have available the same expert statement as considered in Chapter 3: that is, the probability of occurrence of a catastrophic class (CA) scenario (i.e., $T_{\max} \in [300, 500)$ is greater than 0.2, and the probability of occurrence of a scenario of catastrophic (CA) or critical (CR) class ($T_{\max} \in [200, 500)$) is greater than 0.5. Contrary to Chapter 4, however, we also assume that the probability of occurrence of a scenario of safe class ($T_{\max} \in [100, 300)$) is between 0.1 and 0.5. Note that our knowledge of the scenarios concerns three sets (classes): "critical,"

Uncertainty in Risk Assessment: The Representation and Treatment of Uncertainties by Probabilistic and Non-Probabilistic Methods, First Edition. Terje Aven, Piero Baraldi, Roger Flage and Enrico Zio.
© 2014 John Wiley & Sons, Ltd. Published 2014 by John Wiley & Sons, Ltd.

"critical or catastrophic," and "safe," which are not nested (since "safe" is not contained in and does not contain the two other sets). For this reason, this situation of available knowledge cannot be properly treated by possibility theory. Furthermore, as in the previous cases, we are not willing to assign a single value to the probability of the different events and, thus, we cannot use a probabilistic approach. The lack of the nestedness property in the present formulation of the example compared to the examples in Chapter 3 and Chapter 4 is illustrated in Figure 5.1.

To illustrate the idea of obtaining degrees of belief for one question from subjective probabilities for related questions, consider Example 5.2.

According to this example an adequate summary of the impact of evidence includes support of the evidence in favor of the event A "the system has failed" (0.9) and support of the evidence against (0). The former support is reflected in a belief measure, $Bel(A)$, which measures the degree of belief that A will occur, and the latter support is reflected by the plausibility measure, $Pl(A)$, which measures the extent to which evidence does not support $\overline{A}$. The relation between plausibility and belief is

$$Pl(A) = 1 - Bel(\overline{A}).$$ (5.1)

In Example 5.2, the event A refers to "the system has failed," and $Bel(A) = 0.9$ and $Pl(A) = 1$. Note that in evidence theory, as in possibility theory, there are two measures of the likelihood of an event, belief and plausibility. Furthermore, belief and plausibility measures correspond to necessity and possibility measures in the possibility theory.

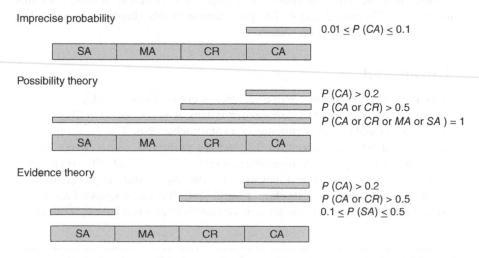

Figure 5.1   Illustration of probability assignments available in Chapter 3 (Imprecise probability), Chapter 4 (Possibility theory), and this chapter (Evidence theory).

Example 5.2

Suppose that a diagnostic model is available which indicates with a reliability (here understood as the frequentist probability of providing the correct result) of 0.9 when a given system fails. Considering a case in which the model does indeed indicate that the system has failed, this fact justifies a 0.9 degree of belief on such event (which is different from the related event of model correctness for which the probability value of 0.9 is available), but only a 0 degree of belief (not 0.1) on the event that the system has not failed. This latter belief does not mean that it is certain that the system has not failed, as a zero probability would imply; it merely means that the model indication provides no evidence to support the fact that the system has not failed. The pair of values $\{0.9, 0\}$ constitutes a belief function on the events "the system has failed" and "the system has not failed." Note that there is no requirement that the belief not committed to the event "the system has failed" $(1 - 0.9 = 0.1)$ must be committed to its negation. Thus, the total allocation of belief can vary to suit the extent of knowledge of the decision maker.

For a formal introduction to the Dempster–Shafer theory of evidence, let Y be an uncertain quantity whose k possible alternative outcomes are $y_1, \ldots, y_k$. The set $S = \{y_1, \ldots, y_k\}$ will be referred to as the sample space. The theory of evidence makes basic belief assignments (BBAs), denoted $m(A)$, for all events A in the power set 2^S of S, that is, on all sets of combinations of events. Hence if $k = 2$, the power set is equal to $\{\varnothing, y_1, y_2, S\}$, where $\varnothing$ is the empty set, as is also explained in Example 5.3 below.

The elements of the power set 2^S of S are referred to as focal sets. The BBA gives values in $[0, 1]$, $m(0) = 0$, and the sum of $m(A)$ over all focal sets is 1, that is,

$$\sum_{A \in 2^S} m(A) = 1. \tag{5.2}$$

Example 5.2 (continued)

The sample space is formed by the two disjoint events "functioning" and "failed," whereas its power set is made up by the following subsets: the empty set (neither "functioning" nor "failed"), "functioning," "failed," and "unknown" ("functioning" or "failed"). The empty set represents a contradiction which is never true as the system must be in a given state at all times; the "unknown" subset represents the situation in which the system may be in either state, in the sense that the available evidence does not allow the exclusion of one or the other. On the basis of the available knowledge, the BBAs in the second column of Table 5.1 are assigned to the elements of the power set.

Table 5.1   Mass, belief, and plausibility for the success or failed state of a system, based on the indication of a fault by a diagnostic model with 0.9 reliability.

Set A	BBA $m(A)$	Belief $Bel(A)$	Plausibility $Pl(A)$
Empty (neither success nor failed)	0	0	0
Functioning	0	0	0.1
Failed	0.9	0.9	1
Unknown (success or failed)	0.1	1	1

Note from (5.2) that it is not required that $m(S) = 1$, or that $m(A) \leq m(B)$ when $A \subseteq B$, or that there be any relationship between $m(A)$ and $m(\overline{A})$. We can see here the contrast to probability theory where we assign probabilities to the single elements/events y_1 and y_2, not the combination "y_1 or y_2" as in evidence theory. In probability theory we compute the probability of the combination "y_1 or y_2" based on the assignments on the single element/event level.

As mentioned above, the value of $m(A)$ pertains solely to the set A and does not imply any additional claim regarding subsets of A; if there is additional evidence supporting the claim that the element y belongs to a subset of A, say $B \subseteq A$, it must be expressed by another BBA on B, that is, $m(B)$.

From the basic probability assignment, the belief and plausibility measures can be obtained as

$$Bel(A) = \sum_{E \subseteq A} m(E), \qquad (5.3)$$

$$Pl(A) = \sum_{E \cap A \neq \varnothing} m(E). \qquad (5.4)$$

The belief in event A is quantified as the sum of masses assigned to all sets enclosed by it; hence, it can be interpreted as a lower bound representing the amount of belief that supports the event. The plausibility of event A is, instead, the sum of the

Example 5.2 (continued)

Table 5.2 reports the values of belief and plausibility for the $2^2 = 4$ possible sets of the power set. Notice that the belief in both the "functioning" and "failed" propositions matches their corresponding mass assignments, because these events have no subsets. Further, the "unknown" event (either "success" or "failed") always has, by definition, belief and plausibility 1. The couple $[Bel, Pl]$ represents the uncertainty on the occurrence of the event, based on the available evidence.

Example 5.1 (continued)

The sample space is formed by the four different criticality classes of the failure scenario, $S = \{SA, MA, CR, CA\}$, whereas its power set is constituted by 16 subsets:

$$2^S = \{\emptyset, SA, MA, CR, CA, SA + MA, SA + CR, SA + CA, MA$$
$$+ CR, MA + CA, CR + CA, SA + MA + CR, SA + MA + CA, SA$$
$$+ CR + CA, MA + CR + CA, S\}$$

In this case the available information is constituted by

$$Bel(CA) = 0.2, Bel(CR + CA) = 0.3, Bel(SA)6 = 0.1,$$
$$Pl(SA) = 1 - Bel(MA + CR + CA) = 0.5.$$

Using (5.5) one obtains the BBAs reported in Table 5.2. Considering, for example, the computation of the BBA of the set $\{CR + CA\}$, one has to consider two contributions to the sum: the belief measure of the set $\{CR + CA\}$ and the set $\{CA\}$. The cardinality of $\{CR + CA\} - \{CR - CA\}$ is equal to 0, and that of $\{CR + CA\} - \{CA\}$ is equal to 1, hence we obtain

$$m(CR + CA) = Bel(CR + CA) - Bel(CA) = 0.5 - 0.2 = 0.3.$$

The BBA of the "unknown" event $\{SA + MA + CR + CA\}$ is obtained by considering that the sum of the BBAs of all the sets of the power set should be 1 (cf. (5.2)).

BBA assigned to all sets whose intersection with A is not empty; hence, it is an upper bound on the probability that the event occurs.

The belief and plausibility measures are often interpreted as probability bounds, in line with imprecise probabilities described in Chapter 3. According to this view, the assigner states that his or her degree of belief is reflected by a probability greater than a comparable urn chance of $Bel(A)$ and less than an urn chance of $Pl(A)$. The analyst is not willing to make a more precise assignment.

Table 5.2   BBA values assigned to the focal sets in Example 5.4.

Set A	BBA $m(A)$
CA	0.2
CR + CA	0.3
SA	0.1
SA + MA + CR + CA	0.4

Table 5.3   Belief and plausibility measures of all 16 subsets of the power set in Example 5.5.

Set A	Belief $Bel(A)$	Plausibility $Pl(A)$
Empty	0	0
SA	0.1	0.5
MA	0	0.4
CR	0	0.7
CA	0	0.9
SA + MA	0.1	0.5
SA + CR	0.1	0.8
SA + CA	0.3	1
MA + CR	0	0.7
MA + CA	0.2	0.9
CR + CA	0.5	0.9
SA + MA + CR	0.1	0.8
SA + MA + CA	0.3	1
SA + CR + CA	0.6	1
MA + CR + CA	0.5	0.9
SA + MA + CR + CA	1	1

Furthermore, notice from the knowledge of the belief and plausibility distributions that it is possible to obtain the basic probability assignment $m(A)$ by

$$m(A) = \sum_{B \subseteq A} (-1)^{card(A-B)} Bel(B), \qquad (5.5)$$

where $card(A)$ expresses the number of elements in the set A. Thus, if for example $A - B$ contains two elements, $Bel(B)$ is added, whereas if $A - B$ contains one element, it is subtracted.

The belief and plausibility measures of all 16 subsets of the power set of $S = \{SA, MA, CR, CA\}$ computed through (5.3) and (5.4) from the BBAs of Table 5.2 are reported in Table 5.3. Notice that, due to the BBA assigned to the safe (SA) scenario, the plausibilities of the events "CA" and "CR or CA" are less than 1.

6

Methods of uncertainty propagation

Once uncertainty has been represented using the approaches described in the previous chapters, it must be propagated through the models used in the risk assessment.

In this chapter, we illustrate the ideas of uncertainty propagation by introducing a generic model $g(X)$ which receives as input a vector $X = (X_1, X_2, \ldots, X_N)$ of N uncertain quantities and provides as output the quantity of interest Z, that is,

$$Z = g(X). \tag{6.1}$$

The uncertainty analysis of Z requires an assessment of the uncertainties about X, and their propagation through the model g to produce an assessment of the uncertainties concerning Z (see Figure 1.6 in Part I). Uncertainty related to the model structure g, that is, uncertainty about the error $Z - g(X)$, is not treated in this book.

Given the different methods to represent the uncertainty of the input quantities (as introduced in Chapters 2–5), there exist different methods for uncertainty propagation. Depending on the type of uncertainty affecting the model input quantities, methods for uncertainty propagation can be classified into level 1 and level 2 (Limbourg and de Rocquigny, 2010; Pedroni and Zio, 2012). The definition of the level 1 and 2 uncertainty propagation settings will be provided in the framework of frequentist probabilities considering the distinction between aleatory and epistemic uncertainty. Although in a Bayesian subjective probability framework all uncertainty is epistemic, the definitions can be easily extended to consider the concept of chance which captures aleatory uncertainty.

For **a level 1 uncertainty propagation setting**, the input quantities are divided into a group, $X_1, \ldots, X_n$, subject to aleatory uncertainty and a group $X_{n+1}, \ldots, X_N$,

Uncertainty in Risk Assessment: The Representation and Treatment of Uncertainties by Probabilistic and Non-Probabilistic Methods, First Edition. Terje Aven, Piero Baraldi, Roger Flage and Enrico Zio.
© 2014 John Wiley & Sons, Ltd. Published 2014 by John Wiley & Sons, Ltd.

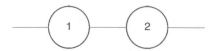

Figure 6.1   Series system.

Example 6.1

We consider a series system comprising two independent components, as in Figure 6.1. The failure time of component 1 (2), denoted T_1 (T_2), is subject to aleatory uncertainty which is represented by the frequentist probability distribution

$$F_1(t_1) = P_f(T_1 \leq t_1)(F_2(t_2) = P_f(T_2 \leq t_2)).$$

The quantity of interest is the failure time of the entire system, T, which is a function of the failure times of the first and second components:

$$T = \min(T_1, T_2). \tag{6.2}$$

Using the notation introduced at the beginning of the chapter, we have $Z = T$, $X = (X_1, X_2)$, and g equal to the "min" function. In this case all the quantities involved are subject to aleatory uncertainty, that is, $n = N$ according to the set-up described at the beginning of the chapter. As a function of the uncertain input quantities, T is itself an uncertain quantity. The uncertainty propagation will provide the frequentist probability distribution $F(t) = P_f(T \leq t)$. Since both input quantities are subject to aleatory uncertainty, this is a typical uncertainty propagation problem in a level 1 setting.

Example 6.2

We consider a structure S which is randomly degrading in time according to a stochastic fatigue crack growth model. The level of degradation of the structure at a future time t is indicated by the unknown quantity $D(t)$ (Figure 6.2). The structure is designed to fulfill its function for a certain time window, called mission time and denoted t_{miss}, and it is considered to be in a failed state when the value of the degradation level $D(t)$ exceeds a certain threshold, D_{max}, whose true value is not exactly known. Let $Y(t)$ denote the state of the structure at time t; that is, define $Y(t) = I(D(t) \leq D_{max})$. The quantity of interest is the state of the structure at the mission time, that is, $Y(t_{miss})$. Note that the state of the structure at mission time is a function of two uncertain input quantities: the crack depth at mission time, $D(t_{miss})$, and the degradation threshold D_{max}. In this example, $D(t_{miss})$ is subject to aleatory uncertainty, since it depends on the stochastic degradation process, whereas the second quantity, D_{max}, is subject to epistemic uncertainty. Since these uncertain quantities are not separated into two hierarchical levels, the propagation of their uncertainties onto the state of the structure at mission time is an example of level 1 uncertainty propagation setting.

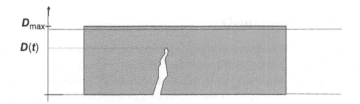

Figure 6.2   Example of crack growth in a degrading structure. The quantity $D(t)$
indicates the crack depth at time t, *and* D_{max} *is the failure threshold; if the crack depth
exceeds* D_{max} *the structure fails.*

subject to epistemic uncertainty, $1 \leq n \leq N$. The level 1 uncertainty propagation
setting assumes that the aleatory quantities, $X_1, \ldots, X_n$, are not subject to any form
of epistemic uncertainty; that is, we know perfectly their probability distributions
(including their parameter values). The group of epistemic quantities, $X_{n+1}, \ldots, X_N$,
is not subject to any form of intrinsic random variation, but only to uncertainty caused
by lack of knowledge: that is, if they were perfectly known they would be represented
by point values.

A **level 2 uncertainty propagation setting** applies if the outcomes of the
input quantities X are subject to aleatory uncertainty described by frequentist
probabilities with parameters Θ subject to epistemic uncertainty. In the case
of "perfect" information which removes all the epistemic uncertainty, we are
back to the level 1 setting. Thus, in a level 2 setting, we assume the presence
of N quantities $X_1, \ldots, X_N$ whose uncertainty is characterized by frequentist
probability distributions $F_i(x_i|\Theta_i) = P_f(X_i \leq x_i|\Theta_i)$, $i = 1, \ldots, N$, where Θ_i is
a vector of the (unknown) parameters of the corresponding probability
distribution.

6.1   Level 1 uncertainty propagation setting

Depending on the methods used to represent the uncertainty on the model input
quantities X, different methods for uncertainty propagation are embraced in a level 1
setting. In this section we consider the following three situations:

- A purely probabilistic framework (cf. Chapter 2), that is, the uncertainty of all
  the model input quantities is represented by probability distributions.

- A purely possibilistic framework (cf. Chapter 4), that is, the uncertainty of all
  model input quantities is represented by possibility distributions.

- A hybrid probabilistic–possibilistic framework, that is, the uncertainty of
  some model input quantities is represented by probability distributions
  and the uncertainty of other model input quantities by possibility
  distributions.

Example 6.2 (continued)

We consider a simple fault tree comprising two independent basic events, B_1 and B_2, linked to the top event A through an OR gate (Figure 6.3).

The quantity of interest is the frequentist probability (or chance) p of the top event, and we want to establish the subjective probability of A, determined as $P(A) = E[p]$. The aleatory uncertainty related to the occurrence of the top event is captured by its frequentist probability (or chance), p, which is a function of the frequentist probabilities (chances), q_1 and q_2, of the basic events B_1 and B_2 as follows:

$$p(q) = q_1 + q_2 - q_1 q_2. \tag{6.3}$$

We assume that the true values of the quantities q_i, $i = 1, 2$, are unknown and that the epistemic uncertainty related to them is described by the subjective probability distribution $H(q') = P(q_1 \leq q_1', q_2 \leq q_2')$. Thus, a level 2 uncertainty propagation setting applies for the assessment of the aleatory uncertainty on the occurrence of the top event, represented by its chance, p. Once the subjective probability distribution, $H(q')$ has been determined, we can compute $P(A)$ from

$$P(A) = E[p] = \int p(q') dH(q'). \tag{6.4}$$

6.1.1 Level 1 purely probabilistic framework

The uncertainty propagation problem in a level 1 setting is addressed here by considering a special case in which all the model input quantities are subject to aleatory uncertainty and a frequentist interpretation of probability is considered. The objective is to determine the probability distribution of the model output $Z = g(X)$ which depends on the joint probability distribution $P_f(X_1 \leq x_1, X_2 \leq x_2, \ldots, X_N \leq x_N)$ of the model input quantities.

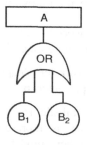

Figure 6.3   Simple fault tree.

Example 6.3

We consider the problem of assessing the performance of a maintenance policy applied to a component of an industrial plant. We assume we know the lifetime model of the component, which encodes two kinds of aleatory input variables: the failure times, T, and the repair times, R (Figure 6.4).

The quantity of interest is the portion D of the mission time, t_{miss}, in which the component is unavailable. The downtime D is a function of the failure and repair times $T = (T_1, T_2, \ldots)$ and $R = (R_1, R_2, \ldots)$ with T_k indicating the time between the end of repair $k - 1$ and the kth failure, and R_k the duration of the kth repair,

$$D = g(T, R). \qquad (6.5)$$

The frequentist probability distributions of T_k and R_k are exponential distributions with uncertain parameters λ and μ, respectively. Since the parameters are subject to epistemic uncertainty, the objective of a level 2 uncertainty propagation setting is to hierarchically propagate the uncertainties onto the model output D.

Exact analytical methods for the propagation of uncertainty are usually not available, except for some simple cases such as a linear combination of normal quantities (Springler, 1979). However, there are a variety of approximate analytical techniques based on a Taylor series expansion of the function g (Cheney, 1966). They typically propagate uncertainty by taking into account moments of the probability distributions of the input quantities, such as the mean, variance, and moments of higher orders, to provide moments of the output quantities. Alternatively, in those cases where the entire probability distribution of the output quantity is required, or the model g is highly non-linear, sampling methods such as Monte Carlo simulation are usually preferred (Morgan and Henrion, 1990; Zio, 2013). When the input quantities are independent and their uncertainty described by the probability density functions $f_i(x_i)$, $i = 1, \ldots, N$, a Monte Carlo simulation requires drawing at random a value

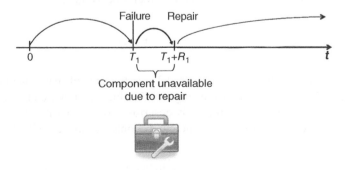

Figure 6.4   Scheme of the failure and repair process of a component.

for each input quantity from its distribution. The obtained set of random values is used as input to the model, computing the corresponding output value. The entire process is repeated M times, where M is a large number, producing M output values which constitute a random sample from the probability distribution of the output quantity Z. In practice, the main steps of the Monte Carlo procedure for uncertainty propagation in a level 1 setting are:

1. Set $j = 0$.

2. Sample the jth set of random realizations $x_1^j, \ldots, x_N^j$ of the input quantities $X_1, \ldots, X_N$ from the probability distributions $f_1(x_1), \ldots, f_N(x_N)$.

3. Calculate the corresponding output values $z^j = g(x_1^j, \ldots, x_N^j)$.

4. If $j < M$ set $j = j+1$ and return to step 1, otherwise build the empirical cumulative distribution $F(z)$ from the samples $z^1, \ldots, z^M$.

Methods to decide the number M of Monte Carlo runs (samples) to be performed in order to achieve a desired degree of accuracy on the estimate of $F(z)$ can be found in Morgan and Henrion (1990). Readers interested in advanced sampling methods which can increase the accuracy of the estimate over traditional Monte Carlo using a limited number of model runs are referred to Zio (2013), whereas methods for sampling in the case of dependence between input variables can be found in Morgan and Henrion (1990). The same uncertainty propagation procedure can be applied in the case where all the input quantities are subject to epistemic uncertainty and their uncertainty is described using subjective probabilities.

6.1.2   Level 1 purely possibilistic framework

Uncertainty propagation in a purely possibilistic framework is typically performed by relying on the extension principle of fuzzy set theory. We first introduce the principle in the case where both the input quantity X, here assumed to be single valued, and output quantity Z of the model g take values on the real line. Assuming that the uncertainty on X is described by the possibility distribution $\pi_X(x)$, the principle allows the function g to be extended to a function that maps from and to the class of all possibility distributions defined on the real line (Zadeh, 1975; Scheerlinck, Vernieuwe, and De Baets, 2012):

$$\pi_Z(z) = \sup_{x, g(x) = z} (\pi_X(x)). \tag{6.6}$$

In the case where the input to the model g is a vector of real-valued quantities $X_1, \ldots, X_N$ and the model output Z a single real-valued quantity, and assuming that the uncertainty on the input quantities is described by the possibility distributions $\pi_1(x_1), \ldots, \pi_N(x_N)$, the extension principle becomes (Zadeh, 1975; Scheerlinck, Vernieuwe, and De Baets, 2012)

$$\pi_Z(z) = \sup_{X_1, \ldots, X_N, g(X_1, \ldots, X_N) = Z} \min\{\pi_{X_1}(x_1), \ldots, \pi_{X_N}(x_N)\}. \tag{6.7}$$

Example 6.3 (continued)

For the chemical reactor example, we consider a physical model g used for evaluating the consequence of a catastrophic failure event. We assume that the model specifies that the consequences C (in arbitrary units) of a catastrophic failure are a quadratic function of the quantity of toxic gas released R (in arbitrary units), that is,

$$C = g(R) = R^2. \tag{6.8}$$

The epistemic uncertainty related to R is described by the triangular possibility distribution:

$$\pi_R(r) = \begin{cases} 0 & \text{if } r < 0 \text{ or } r > 2 \\ r & \text{if } 0 \le r \le 1 \\ 2 - r & \text{if } 1 < r \le 2, \end{cases} \tag{6.9}$$

as shown in Figure 6.5.

In this case, the extension principle provides the possibility distribution describing the uncertainty to which the model output C is subject. According to (6.6), we first fix a value $c > 0$ for the consequences and find the quantity of toxic gas released r such that $c - r^2$. In this particular case, the solutions are $r_1 = -\sqrt{c}$ and $r_2 = \sqrt{c}$. We then evaluate the possibility distribution of the quantity of toxic gas released, π_R, at the identified r values, that is, $\pi_R(r_1)$ and $\pi_R(r_2)$, and assign to the possibility distribution of the consequence, π_C, the maximum of $\pi_R(r_1)$ and $\pi_R(r_2)$. Since, in this case, $\pi_R(r_1) = \pi_R(-\sqrt{c})$ is equal to 0, the maximum value is obtained as $\pi_R(r_1) = \pi_R(\sqrt{c})$ which, according to the definition of $\pi_R(r)$ in (6.9), is given by

$$\pi_C(c) = \pi_R(r_2) = \pi_R(\sqrt{c}) = \begin{cases} 0 & \text{if } c \le 0 \\ \sqrt{c} & \text{if } 0 < c \le 1 \\ 2 - \sqrt{c} & \text{if } 1 < c \le 4 \\ 0 & \text{if } c > 4. \end{cases} \tag{6.10}$$

The resulting possibility distribution of C is shown in Figure 6.6.

Note that, as expected, the most unsurprising and usual value of the model output, that is, the catastrophic failure consequence characterized by a possibility distribution equal to 1, is obtained for a quantity of toxic gas released equal to 1, which is the most unsurprising and usual value of the model input, that is, the gas released value associated with a possibility distribution of 1.

The use of the minimum operator to combine the possibility distributions is justified by the fact that the joint possibility distributions of the N input quantities, $\pi_{X_1,\ldots,X_N}(x_1,\ldots,x_N)$, is defined by the minimum of the possibility distributions $\pi^{X_1}(x_1),\ldots,\pi^{X_N}(x_N)$.

An alternative formulation of the extension principle which has been shown to be equivalent to the one in (6.7) is based on the representation of the output possibility

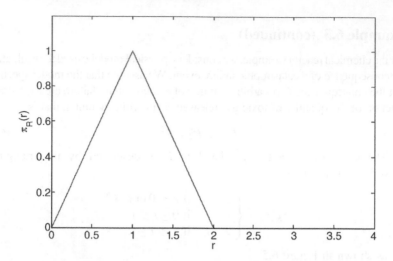

Figure 6.5   Possibility distribution describing the uncertainty of the quantity of toxic gas released.

distribution in the form of a nested set of intervals $A_\alpha = [\underline{z}_\alpha, \overline{z}_\alpha] = \{z : \pi_Z(z) \geq \alpha\}$, which are usually referred to as α-cuts (Section 4.1). Indicating as $X_{1\alpha}, \ldots, X_{N\alpha}$ the N_α α-cuts of the input quantities $X_1, \ldots, X_N$, the extension principle, for a given value of α in $[0, 1]$, becomes

$$\underline{z}_\alpha = \inf(g(x_1, \ldots, x_N), x_1 \in X_{1\alpha}, \ldots, x_N \in X_{N\alpha}),$$
$$\overline{z}_\alpha = \sup(g(x_1, \ldots, x_N), x_1 \in X_{1\alpha}, \ldots, x_N \in X_{N\alpha}). \qquad (6.11)$$

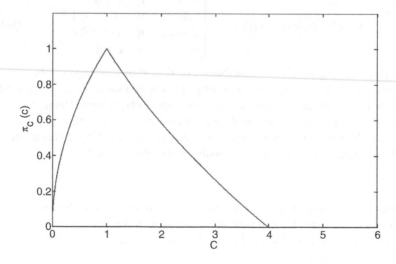

Figure 6.6   Possibility distribution of the failure consequence obtained by applying the fuzzy extension principle.

Example 6.3 (continued)

For the chemical reactor example, we now consider an alternative model for evaluating the consequence of a catastrophic failure event. We now assume that the quantity of toxic gas released, R, depends on the quantities S_1 and S_2 of two different types of molecules (say A and B) inside the reactor:

$$R = g(S) = S_1 + S_2. \tag{6.12}$$

The epistemic uncertainty on S_1 and S_2 is described by the triangular possibility distribution shown in Figure 6.7 and defined for both $i = 1$ and $i = 2$ by

$$\pi_{S_i}(s_i) = \begin{cases} 0 & \text{if } s_i < 0 \text{ or } s_i > 1 \\ 2s_i & \text{if } 0 < s_i \leq 0.5 \\ 2(1 - s_i) & \text{if } 0.5 < s_i \leq 1. \end{cases} \tag{6.13}$$

According to the extension principle, we can propagate the uncertainty from S_1 and S_2 to R. Consider, as an example, the computation of the possibility distribution of R, $\pi_R(r)$, for $r = 1$. Then, in order to satisfy the condition of (6.7) that $g(X_1, \ldots, X_N) = Z$, which in this case becomes $s_1 + s_2 = 1$, we consider all the pairs of values $(s_1, s_2) = (1 - \varepsilon, \varepsilon)$ with ε in the interval $[0, 1]$. Then, for $0 \leq \varepsilon < 0.5$ we obtain $\min(\pi_{S_1}(1 - \varepsilon), \pi_{S_2}(\varepsilon)) = \min(2\varepsilon, 2\varepsilon) = 2\varepsilon$ and for $0.5 < \varepsilon \leq 1$ we obtain $\min(\pi_{S_1}(1 - \varepsilon), \pi_{S_2}(\varepsilon)) = \min(2 - 2\varepsilon, 2 - 2\varepsilon) = 2 - 2\varepsilon$. Note that the most unsurprising and usual value of the model output, that is, the toxic gas released such that $\pi_R(r) = 1$, is obtained by computing the model $g(S)$ in correspondence to the two most unsurprising and usual values of the model input quantities, that is, quantities of the two molecules equal to 0.5. The possibility distribution of the release of toxic gas obtained by applying the fuzzy extension principle for all values of $s > 0$ is the same possibility distribution previously reported in Figure 6.5.

According to this latter formulation, the extension principle is equivalent to performing interval analysis with α-cuts and, as a result, imposes an assumption about a strong dependence between the information sources supplying the input possibility distributions. In practice, it has been observed that its application can be questionable in the case where different experts provide the possibility distributions describing the uncertainty on different input quantities. The reader interested in this topic may refer to Pedroni and Zio (2012) and Baudrit, Dubois, and Guyonnet (2006).

6.1.3   Level 1 hybrid probabilistic–possibilistic framework

We now assume that the uncertainty related to the first n input quantities, $X_1, \ldots, X_n$, $n < N$, is described using the probability distributions $F_1(x_1), \ldots, F_n(x_n)$, whereas the uncertainty related to the remaining $N - n$ quantities $X_{n+1}, \ldots, X_N$ is represented using the possibility distributions $\pi_{n+1}(x_{n+1}), \ldots, \pi_N(x_N)$.

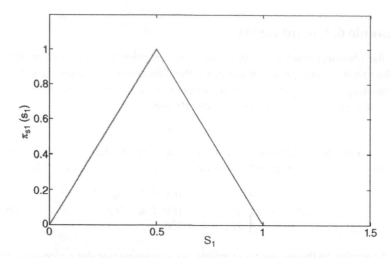

Figure 6.7 Possibility distribution describing the uncertainty on the quantity of molecules A *and* B *inside the reactor.*

The propagation of the hybrid uncertainty information through the function g can be performed by combining Monte Carlo sampling with the extension principle (see Figure 6.8). The main steps of the procedure are:

- Repeated Monte Carlo sampling of the probabilistic quantities.

- Application of the extension principle of fuzzy set theory (cf. Section 6.1.2) to process the uncertainty associated with the possibilistic quantities.

More specifically, for a vector of fixed values of the quantities $x_1^j, \ldots, x_n^j$ obtained in the jth Monte Carlo sampling, the extension principle defines the corresponding possibility distribution of the output Z as

$$\pi^j(z) = \sup_{x_{n+1},\ldots,x_N\,:\,g(x_1^j,\ldots,x_n^j,x_{n+1},\ldots,x_N)=z} \left(\min(\pi_{n+1}(x_{n+1}),\ldots,\pi_N(x_N)) \right). \qquad (6.14)$$

Example 6.3 (continued)

According to (6.11), uncertainty propagation through the function $R = S_1 + S_2$ can be performed by considering the α-cuts. Operationally, one can fix a value of α, identify the corresponding α-cuts, $S_{1\alpha}$ and $S_{2\alpha}$, of the quantities of S_1 and S_2, and compute $\underline{z}_\alpha$ and $\overline{z}_\alpha$ from (6.11). Considering, for example, $\alpha = 0.5$, we obtain $S_{1\alpha} = S_{2\alpha} = [0.25, 0.75]$. Then, the minimum of $S_1 + S_2$ with S_1 and S_2 varying in $[0.25, 0.75]$ is 0.5, which is obtained for $s_1 = s_2 = 0.25$, and the maximum is 1.5 obtained for $s_1 = s_2 = 0.75$. Thus, the α-cut of 0.5 of R is $[0.5, 1.5]$.

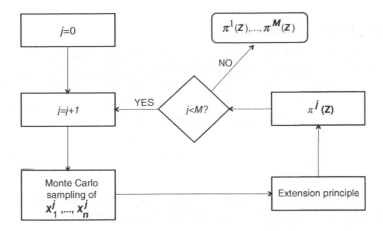

Figure 6.8   Scheme of the hybrid probabilistic–possibilistic uncertain propagation method.

Introducing the α-cut of the output quantity $Z_\alpha = [\underline{z}_\alpha, \bar{z}_\alpha]$ and of the ith input quantity $X_{i\alpha}$, $i = n+1, \ldots, N$, the extension principle can also be formulated as in (6.11):

$$\underline{z}_\alpha = \inf x_{n+1} \in X_{(n+1)\alpha}, \ldots, X_N \in X_{N\alpha}, (g(x_1^j, \ldots, x_n^j, x_{n+1}^j \ldots, x_N^j)$$
$$\bar{z}_\alpha = \sup x_{n+1} \in X_{(n+1)\alpha}, \ldots, X_N \in X_{N\alpha}, (g(x_1^j, \ldots, x_n^j, x_{n+1}^j, \ldots, x_N^j)). \tag{6.15}$$

At the end of the procedure, having Monte Carlo-sampled M values of the probabilistic quantities, an ensemble of realizations of possibility distributions is obtained, that is, a set of possibility distributions $(\pi^1, \ldots, \pi^M)$. Then, using (4.1) and (4.2) for each set A contained in the domain of the output variable Z, it is possible to obtain the corresponding possibility measure $\Pi^j(A)$ and necessity measure $N^j(A)$ as

$$\Pi^j(A) = \max_{z \in A}\{\pi^j(z)\}, \tag{6.16}$$

$$N^j(A) = \min_{z \in A}\{1 - \pi^j(z)\} = 1 - \Pi^j(\bar{A}). \tag{6.17}$$

The M different realizations of possibility and necessity measures can then be combined to obtain the belief $Bel(A)$ and the plausibility $Pl(A)$, respectively, for any set A (Baudrit, Dubois, and Guyonnet, 2006) as

$$Bel(A) = \frac{1}{M}\sum_{j=1}^{M} N^j(A) \tag{6.18}$$

$$Pl(A) = \frac{1}{M}\sum_{j=1}^{m} \Pi^j(A). \tag{6.19}$$

Example 6.3 (continued)

For the chemical reactor example, we consider the same model as was used in Section 6.1.2 for evaluating the quantity of toxic gas released, R, which depends on the quantities S_1 and S_2 of the two chemical molecules inside the reactor, according to (6.7). As in Section 6.1.2 we assume that the epistemic uncertainty on S_1 is described by the triangular possibility distribution

$$\pi_1(s_1) = \begin{cases} 0 & \text{if } s_1 < 0 \text{ or } s_1 > 1 \\ 2s_1 & \text{if } 0 < s_1 \le 0.5 \\ 2(1-s_1) & \text{if } 0.5 < s_1 \le 1, \end{cases} \tag{6.20}$$

whereas, differently from the previous case, the epistemic uncertainty on S_2 is described by a beta probability density function with parameter values $\beta_1 = 5$ and $\beta_2 = 20$ (Figure 6.9):

$$p_2(s_2) = \frac{\Gamma(\beta_1)}{\Gamma(\beta_1)+\Gamma(\beta_2)} (s_2)^{\beta_1-1}(1-s_2)^{\beta_2-1}.$$

The uncertainty propagation procedure requires Monte Carlo sampling of a value of S_2 from its beta probability distribution and then applying the fuzzy extension principle. Considering, for example, a case in which the Monte Carlo sampling of S_2 produces the value 0.43, we can apply the extension principle in the form of (6.11):

$$\underline{z}_\alpha = \inf s_1 \in S_{1\alpha}(0.43 + s_1)$$
$$\overline{z}_\alpha = \sup s_1 \in S_{1\alpha}(0.43 + s_1, s_1 \in S_{1\alpha}). \tag{6.21}$$

In this case, the smallest and largest values of $0.43 + s_1$ are found in correspondence to the minimum and maximum values of the interval $S_{1\alpha}$, that is, $\alpha/2$ and $1 - \alpha/2$, respectively:

$$\underline{z}_\alpha = 0.43 + \alpha/2$$
$$\overline{z}_\alpha = 0.43 + 1 - \alpha/2. \tag{6.22}$$

The resulting possibility distribution is shown in Figure 6.10. Figure 6.11 shows the possibility distributions obtained for 100 different sampled values of S_1.

Equations (6.18) and (6.19) are often used to compute the belief $Bel([0, z))$ and the plausibility $Pl([0, z))$, which are then interpreted as bounding cumulative distributions of the cumulative distribution of Z, $H(z) = P(Z \le z)$.

It is important to be aware of that, in addition to the assumption of dependence between information sources underlying the extension principle, the hybrid propagation method assumes stochastic independence between the group of probabilistic quantities and the group of possibilistic quantities.

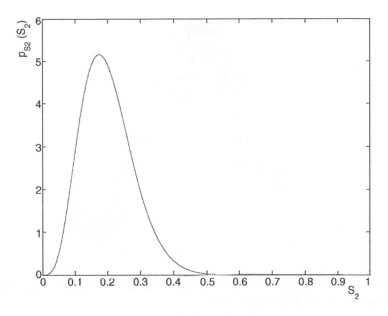

Figure 6.9   Probability density function describing the uncertainty of S_2.

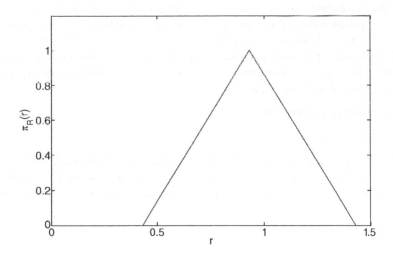

Figure 6.10   Possibility distribution of the quantity of toxic gas released obtained in correspondence to a Monte Carlo sample of the quantity S_2 of molecule B inside the reactor equal to 0.43.

6.2   Level 2 uncertainty propagation setting

In this section, we consider the model set-up $Z = g(X)$, where the N input quantities $X = (X_1, \ldots, X_N)$ are subject to aleatory uncertainty which is represented using the

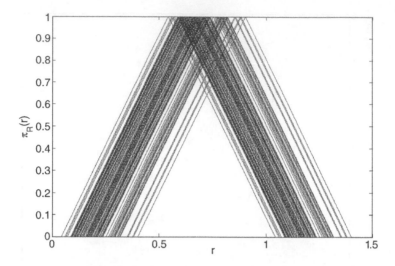

Figure 6.11   Possibility distributions of release of toxic gas in correspondence to 100 Monte Carlo sampled values of S$_1$.

frequentist probability density functions $f_i(x_i|\Theta_i)$, $i = 1, \ldots, N$, where Θ_i is (are) the uncertain parameter(s) of the probability density functions. For the purpose of illustration, we will refer to the particular case in which the probability density functions depend on a single parameter. Furthermore, we assume that the parameters are subject to epistemic uncertainty and we will consider two different methods for the representation of their uncertainty:

- Probability distribution
- Evidence theory.

The former case leads to a purely probabilistic level 2 framework; the latter case results in a hybrid scheme with the uncertainty on X described by probability models with unknown parameters, whose uncertainty is described using evidence theory.

Example 6.3 (continued)

Figure 6.12 shows 100 realizations of the possibility and necessity measures of the set $[0, r)$ corresponding to the possibility distributions shown in Figure 6.11. Finally, Figure 6.13 shows the corresponding belief and plausibility distributions $Bel([0,r))$ and $Pl([0,r))$ which are obtained by applying (6.18) and (6.19) and may be interpreted as bounding cumulative distributions of the quantity of toxic gas released.

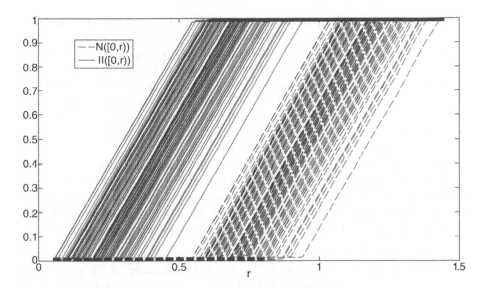

Figure 6.12   Possibility and necessity measures of the set [0, r) corresponding to the possibility distributions shown in Figure 6.11.

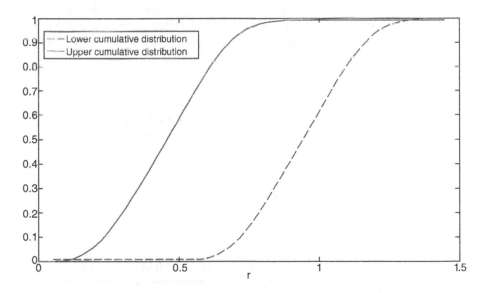

Figure 6.13   Belief and plausibility measures of the set [0, r).

6.2.1   Level 2 purely probabilistic framework

In this framework we assume that both the epistemic uncertainty on the parameters Θ and the aleatory uncertainty on the input quantities $X_1, \ldots, X_N$ are represented using subjective and frequentist probability distributions, respectively. In particular, we let

$h_i(\theta_i)$ denote the subjective probability density function describing the uncertainty of the parameter Θ_i of the frequentist probability density function $f_i(x_i|\Theta_i)$.

In this context, various techniques for uncertainty propagation have been proposed in the literature. As previously described only a few simple cases can be addressed by analytical methods. The most common approach is Monte Carlo simulation. Other techniques which require lower computational efforts, such as those based on local expansion, most probable point, functional expansion, and numerical integration, have been considered (e.g., Xiong et al., 2011), but do not give the same accuracy as Monte Carlo simulation.

In this section we will consider Monte Carlo simulation. The aleatory and epistemic uncertainties are propagated by a two-level (or double-loop) simulation (Cullen and Frey, 1999). In the outer simulation loop the values of the epistemically uncertain parameters are sampled and fed into the inner loop where the aleatory quantities are sampled. In practice, the following two steps are repeated (Figure 6.14):

- Outer loop: Monte Carlo sampling of the epistemically uncertain parameters Θ_i from the probability density $h_i(\theta_i)$. The obtained values at the generic j_eth repetition of this step will be referred to as $\theta^{j_e} = (\theta_1^{j_e}, \ldots, \theta_N^{j_e})$, and the number of repetitions of this step as M_e.

- Inner loop: Repeated Monte Carlo sampling of the quantities $X_1, \ldots, X_N$ from the corresponding probability densities $f_i\left(x_i|\theta_i^{j_e}\right)$ conditioned at the values $\theta_i^{j_e}$ of the epistemically uncertain parameters Θ_i sampled in the inner loop. The number of repetitions of the Monte Carlo sampling in this step is referred to as

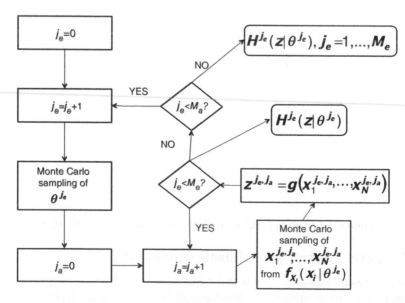

Figure 6.14   Scheme of the Monte Carlo two-loop uncertainty propagation method.

M_a. Alternatively, the inner loop could be performed by using other probabilistic techniques such as the first/second-order reliability method (FORM/SORM) (Kiureghian, Lin, and Hwang, 1987). The output of this step is an empirical cumulative distribution, $H_{j_e}(z|\theta_i^{j_e})$, for the model output Z conditioned on the values of the parameters $\theta_i^{j_e}$.

The application of these two steps provides a set of a cumulative distributions, $H_{j_e}(z|\theta_i^{j_e}), j_e = 1, \ldots, M_e$, one for each repetition of the outer loop. The interpretation of these distributions may not be straightforward in practical terms, given the difficulty of extracting concise information on the output uncertainty from such a representation. A possible way out is to fix a given percentile of the cumulative distributions, for example, the 95th percentile, and then build the probabilistic distribution of this percentile from the different realizations of the epistemic parameters Θ_i. Alternatively, for each sampling of the epistemic parameters in the outer loop, Helton (2005) estimates the corresponding expected values of the output quantities $E[Z|\theta^{j_e}]$ in the inner loop. The overall (unconditional) expected value of Z, $E[E[Z|\Theta]]$, is computed as the mean value of $E[Z|\theta^{j_e}]$ in all the repetitions of the outer loop, whereas a measure of the epistemic uncertainty on Z is provided by the distribution of the obtained $E[Z|\theta^{j_e}]$.

6.2.2   Level 2 hybrid probabilistic–evidence theory framework

As in the previous section, we consider the model set-up with $Z = g(X_1, \ldots, X_N)$ where the uncertainty related to the outcome of the N input aleatory quantities $X_1, \ldots, X_N$ is represented by frequentist probability density functions $f_i(x_i|\Theta_i)$, $i = 1, \ldots, N$. Differently from the purely probabilistic case, however, we describe the epistemic uncertainty on the parameters Θ_i using the framework of evidence theory. In particular, we define focal sets and assign their associated masses for each parameter $\Theta_i, i = 1, \ldots, N$. This information is represented within the evidence space (S_i, J_i, m_i), with S_i indicating the domain of the parameter Θ_i, that is, the set of possible values for the parameters, $J_i = \left\{ J_i^{(l)}, l = 1, \ldots, u_i \right\}$ the set of the u_i focal sets, and $m_i\left(J_i^{(l)}\right)$ the masses associated to the focal set $J_i^{(l)}$.

We present below a hybrid Monte Carlo–evidence theory uncertainty propagation method, which maps the different combinations of uncertain parameters into some selected summarizing measures (mean, quartiles, etc.) of the output quantity (Helton, Johnson, and Oberkampf, 2004).

The first step in the method considers the N-dimensional space of the parameters $\Theta = (\Theta_1, \ldots, \Theta_N)$. According to Helton, Johnson, and Oberkampf (2004), the evidence space (S_Θ, J, m_J) characterizing the uncertainty in this multi-dimensional space is constructed on the basis of the mono-dimensional evidence spaces of the single parameters Θ_i, $i = 1, \ldots, N$. Specifically, S_Θ is the set containing the points $\Theta = [\Theta_1, \ldots, \Theta_N]$ belonging to the Cartesian product of the sample spaces of the N uncertain parameters, that is, $\{\Theta | \Theta = [\Theta_1, \ldots, \Theta_N] \in S_1 \times \cdots \times S_N\}$; the

Example 6.3 (continued)

We suppose that the failure rate λ_1 of the exponential failure time distribution is known to take a value in the interval $[0, 1/100]$ and that three different experts (indexed by $l, l = 1, 2, 3$) have provided the intervals $J_1^{(l)} = \left[a_1^{(l)}, b_1^{(l)}\right], l = 1, 2, 3,$ that they believe to contain the true parameter value (Table 6.1). These assignments are made on the basis of their experience and independently of one another. Analogously, suppose that the repair rate λ_2 of the exponential repair time distribution is known to take a value in the interval $[0, 1]$ and that two other experts (different from those used for the failure rate assignments above) have provided the intervals $J_2^{(l)} = \left[a_2^{(l)}, b_2^{(l)}\right], l = 4, 5,$ that they believe to contain the true parameter value (Table 6.1). All the experts are considered equally credible.

Following the procedure proposed in Helton, Johnson, and Oberkampf (2004), we consider that the focal sets with associated non-null masses are $J_1^{(1)}, J_1^{(2)}, J_1^{(3)},$ and $J_2^{(4)}, J_2^{(5)}$ (Figure 6.15 and Figure 6.16). Coherently with the assumption that all sources of information are equally credible, the BBA associated to a given focal element $J_i^{(l)}$ is the fraction of the sources that specified that focal set. Thus, in our case, the focal elements $J_1^{(1)}, J_1^{(2)}, J_1^{(3)}$ have associated BBAs $m_1\left(J_1^{(1)}\right) = m_1\left(J_1^{(2)}\right) = m_1\left(J_1^{(3)}\right) = 1/3,$ and $J_2^{(4)}, J_2^{(5)}$ have associated BBAs $m\left(J_2^{(4)}\right) = m\left(J_2^{(5)}\right) = 1/2.$

corresponding set of focal elements is $J = \{J_1 \times \cdots \times J_N\}$. Considering the parameters of Θ independently, we associate a BBA m_J to each focal element $E = (E_1, \ldots, E_N)$ of J. In analogy to the case of probability spaces, where the probability of the combination of events pertaining to different spaces is given by the product of the probabilities of the single events, the BBA associated to E is given by

$$m_J(E) = \begin{cases} \prod_{i=1,\ldots,N} m_i(E_i) & \text{if } E = (E_1, \ldots, E_N) \in J \\ 0 & \text{otherwise.} \end{cases} \quad (6.23)$$

Table 6.1   Uncertainty ranges for the parameters independently provided by different experts.

Parameter	Expert 1		Expert 2		Expert 3		Expert 4		Expert 5	
	Min	Max	Min	Max	Min	Max	Min	Max	Min	Max
Θ_1	1/1000	1/500	1/2000	1/750	1/800	1/200				
Θ_2							1/12	1/4	1/5	1/2

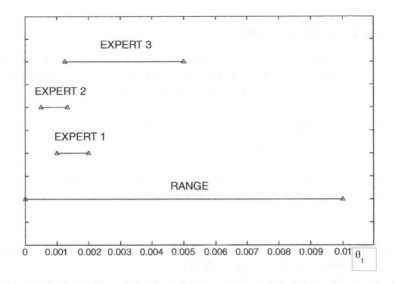

Figure 6.15 Ranges of possible values for parameter Θ_1 provided by three experts.

Note that the independence assumption adopted in this case, usually referred to as "random set independence" (Couso, Moral, and Walley, 1999), comes from the assumption that the experts providing estimates of a parameter are different from those providing estimates of any other parameter. This assumption has been used in Helton, Johnson, and Oberkampf (2004) with respect to the mean and standard deviation of a lognormal distribution. Note, however, that (6.23) cannot be applied to cases in which different forms of independence hold between the parameters; other

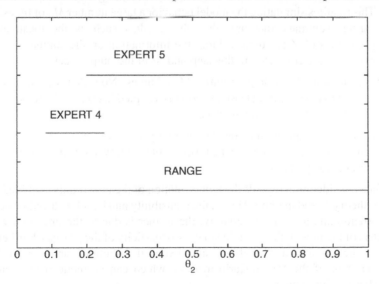

Figure 6.16 Ranges of possible values for parameter Θ_2 provided by two experts.

Example 6.3 (continued)

The evidence space (S_Θ, J, m_J) characterizing the uncertainty in the two-dimensional space of $\Theta = [\Theta_1, \Theta_2]$ is defined by the sample space $S_\Theta = [S_1 \times S_2] = [[0, 1/100] \times [0, 1]]$; the set of the focal sets is

$$J = \{E_1 = J_1^{(1)} \times J_2^{(4)}, \ E_2 = J_1^{(1)} \times J_2^{(5)}, \ E_3 = J_1^{(2)} \times J_2^{(4)}, \ E_4 = J_1^{(2)} \times J_2^{(5)},$$

$$E_5 = J_1^{(3)} \times J_2^{(4)}, E_6 = J_1^{(3)} \times J_2^{(5)}\}$$

and the BBA, m_J, assigned to all six focal sets is $1/3 \times 1/2 = 1/6$, which is the product of the probability masses m_1 and m_2 assigned to the intervals.

formulas have been proposed to address these different situations (Couso, Moral, and Walley, 1999; Su *et al.*, 2011; Yager, 2011).

The methodology for propagating the uncertainties from Θ to the model output Z consists of the following steps:

1. Define a probability distribution $p_\Theta(\theta)$ on S_J. This probability distribution will be used for generating a sample $\theta^j = [\theta_1^j, \ldots, \theta_N^j]$ of Θ.

2. Generate a random sample $\theta^j = [\theta_1^j, \ldots, \theta_N^j]$ from the N-dimensional space Θ, coherently with the distribution defined previously in step 1.

3. Once the values of the parameters of Θ have been fixed, one can perform a standard Monte Carlo propagation of the uncertainty affecting the stochastic variables $X_1, \ldots, X_N$ in order to obtain the uncertainty on the output variable Z. This requires simulating the model behavior a large number M_a of times. Then, various summary measures $\Phi = (\Phi_1, \ldots, \Phi_V)$ such as the mean and the percentiles of Z are used to lump the information on the simulated Z. Such measures are computed in this step and form the output vector.

4. Repeat steps 2–3 a large number M_e of times. Note that M_e must be large enough to ensure that at least one point is sampled in each of the focal elements J of the multi-dimensional space.

5. Estimate the uncertainty on the summary measure $\Phi_v, v = 1, \ldots, V$, of the output Z of the function g in the forms of plausibility and belief distributions in the interval $[-\infty, \phi_v]$.

A final consideration regards the representation of the uncertainty provided by the evidence theory-based method. That is, the plausibility and belief distributions encode both epistemic and aleatory uncertainty: the former is due to the uncertainty in the parameters of the model, the latter due to the stochasticity of the process. Furthermore, note that the computed plausibility and belief distributions are subject to the estimation error of the Monte Carlo method, which can be reduced by increasing the number of simulations.

7

Discussion

While the risk assessment framework based on a probabilistic representation of uncertainties (probabilistic risk assessment, PRA) has proved to be a useful tool in a wide range of applications, a growing number of researchers and analysts are pointing at the limitations of the probability-based approaches for assessing risk in some circumstances. The main point advocated is that the knowledge and information (or lack of such) available for the analysis cannot be properly reflected by probabilities in all situations. To address these situations, approaches other than purely probabilistic ones have been suggested, including those described in the preceding chapters. Much of the development has been of a technical nature, with less emphasis on principles and guidelines for use in practical risk assessment. In this chapter, we provide a discussion raising some concerns and give constructive criticisms on the various approaches proposed. The discussion is taken from or based on Flage et al. (submitted), who identify and discuss five directions of development for uncertainty treatment in the context of risk analysis:

1. Probability

2. Non-probabilistic representations with the interpretation of lower and upper probabilities

3. Non-probabilistic representations with interpretations other than lower and upper probabilities

4. Hybrid approaches for combining probabilistic and non-probabilistic representations

5. Semi-quantitative methods.

In the following we use these five directions as a point of departure for a discussion.

Uncertainty in Risk Assessment: The Representation and Treatment of Uncertainties by Probabilistic and Non-Probabilistic Methods, First Edition. Terje Aven, Piero Baraldi, Roger Flage and Enrico Zio.
© 2014 John Wiley & Sons, Ltd. Published 2014 by John Wiley & Sons, Ltd.

7.1 Probabilistic analysis

Probabilistic analysis is the predominant method used to handle the uncertainties involved in risk analysis, both aleatory and epistemic types. It seems correct to say that probability is indeed perfectly suited to describe aleatory uncertainty, in its limiting relative frequency interpretation. When used as a representation of epistemic uncertainty, the suitable interpretation is the subjective one typical of the Bayesian probabilistic framework (Bedford and Cooke, 2001; NRC, 2009). Several authors argue that the problem of using probability in application lies with the measurement procedures and not with the probability concept in itself (e.g., Lindley, 2006; Bernardo and Smith, 1994; O'Hagan and Oakley, 2004). One development direction then, as suggested by Lindley (2006) and O'Hagan and Oakley (2004), is to retain probability as the sole representation of uncertainty and to focus on improving the measurement procedures for probability. The thesis is that no alternative to probability is needed, as "probability is perfect" (Lindley, 2000; O'Hagan and Oakley, 2004; North, 2010). Supporters of this thesis may acknowledge that there is a problem of imprecision in probability assignments, but this is considered a problem related to the elicitation of probabilities, and not a problem of the probability concept.

This is a strong thesis that in our view is not justifiable as a general statement. We need to clarify the situations addressed, making some distinctions that are important for the practice of risk assessment and management. The classical case of decision analysis for risk management considers the situation in which the assessor of the probabilities and risk is also the decision maker. In this case, it is possible to argue that the use of (subjective) probability as the only uncertainty measure is proper, because the subjectivity of the assessment is brought into the decision scheme coherently by the assessor him- or herself. In simple words, the assessor expresses judgments to arrive at an assessment of the probabilities which he or she uses for the decision making. However, the situation most commonly encountered in risk assessment practice is different, when there is one (or more) assessors who perform the (probabilistic) risk assessment, and whose results are fed to decision makers other than the assessor(s). What characterizes such situations is, specifically, the following:

1. A risk assessment is carried out by a risk analyst/expert group following a request from the decision maker(s) (or other stakeholder(s)). The assessment is dependent on the subjective expertise of the assessor(s).

2. The aim is to carry out the assessment, whose results are independent of the decision maker (and other stakeholders).

3. Several stakeholders other than the decision maker will often be informed about the assessment and its results.

4. The decision maker(s) and stakeholders will perform the decision-making process according to their subjective values and preferences, and be informed by the risk results obtained by the assessor(s) on the basis of his or her (subjective) expertise.

One example is the societal safety issue where politicians are to make decisions on protection measures, informed by the results of a risk assessment characterizing the risks associated with alternative options, and where the risk assessment is performed by expert risk analysts.

It is clear that, in these situations, the decisions taken by the decision makers are influenced also by the knowledge that the assessors put into the risk assessment, through the output probabilities calculated. The strength of knowledge of the system and phenomena studied determines the strength of the assessment and its influence on the decision maker. On the contrary, when the strength of knowledge is poor, this should be transparent and acknowledged in the decision process which relies on the provided probabilities.

In the above scenario of decision making, we advocate strongly that the output that should be retained from the risk assessment and accounted for in the decision-making process comprises two main components:

- The quantitative description given by the uncertainty measure used Q (e.g., probability P).

- The background knowledge K used in the assessment.

We regard this as fundamentally important, because the probabilities of the risk assessment are by definition subjective, or inter-subjective to some extent, depending on the situation analyzed. When a risk assessment is performed, background knowledge is to a large extent expressed in the assumptions made. In the case of relatively poor knowledge, stronger assumptions are made. With strong knowledge, weaker assumptions are required. Table 7.1 presents a matrix of different situations reflecting different "states" of Q and K.

Consider the case (1), which is characterized by strong assumptions and a subjective measure of uncertainty, understood as a subjective probability. By replacing the probability P by an interval probability, the intention is to move the situation to (2), (3), (4), or (5). The idea is to use intervals that are less subjective and

Table 7.1 A matrix reflecting different situations of uncertainty descriptions Q and knowledge K (the numbers are just references for these elements of the matrix).

Uncertainty measure Q	Subjective	Inter-subjectivity among experts	Broad inter-subjectivity
Knowledge K			
Strong – weak assumptions made in the risk assessments		(3)	(5)
Medium – strong assumptions made in the risk assessments		(2)	(4)
Poor – strong assumptions made in the risk assessments	(1)		

based on assumptions that are not so strong. In most cases the change would lead to the situation (2), as the intervals would need also to be based on some assumptions and there are always aspects of the uncertainty description that are not generally inter-subjective.

We stress that no value ranking of the different elements of the matrix is performed in Table 7.1. A subjective assignment (1) may well serve the purpose of the assessment in some cases, where the focus is on reporting some analysts' view on specific issues. The results are acknowledged as subjective but still considered informative for the relevant decision making. In other cases, we may look for more inter-subjective results, more independent of the judgments of specific analysts or experts. Normally, we find that a combination of different approaches can be useful to support the decision making. We need to produce both subjective judgments and beliefs by selected analysts and experts, and we need to produce more inter-subjective results where the knowledge and lack of knowledge available are laid out "plain and flat" with no additional information inserted.

7.2 Lower and upper probabilities

The development of probability as a measurement of uncertainty is based on an axiom that says that precise measurements of uncertainties can be made; see, for example, Bernardo and Smith (1994, p. 31). Being an axiom, this is taken as a fundamental truth and not questioned within the theory or paradigm. Many research-ers have, however, questioned this assumption of, and requirement for, precision (e.g., Dempster, 1967; Walley, 1991). And, indeed, the issue of imprecision is not so easily dismissed, as shown for example by these statements from Bayesian proponents:

> Some writers have considered the axioms carefully and produced objections. A fine critique is Walley (1991), who went on to construct a system that uses a pair of numbers, called upper and lower probabilities, in place of the single proba-bility. The result is a more complicated system. My position is that the com-plication seems unnecessary.
>
> *(Lindley, 2000, p. 298)*

> In practice, there might, in fact, be some interval of indifference [. . .].
> *(Bernardo and Smith, 1994, p. 32)*

> We would not wish, however, to be dogmatic about this. Our basic commitment is to quantitative coherence. The question of whether this should be precise, or allowed to be imprecise, is certainly an open, debatable one, and it might well be argued that "measurement" of beliefs and values is not totally analogous to that of physical "length". [. . .] In this work, we shall proceed on the basis of a pre-scriptive theory which assumes precise quantification, but then pragmatically acknowledges that, in practice, all this should be taken with a large pinch of salt and a great deal of systematic sensitivity analysis.
>
> *(Bernardo and Smith, 1994, p. 99)*

Interval or imprecise probabilities are proposed as an uncertainty measure, alternative to "precise" (single-valued) subjective probabilities. The motivation is that intervals correspond better to the (weak) information available in many cases. The intervals can be elicited by direct arguments or constructed indirectly from assigned possibility functions, or from mass functions as in the framework of evidence theory, as shown in the previous chapters.

Consider the problem of (subjectively) describing the uncertainty of a quantity y which is known to take the value 1, 2, 3, 4, or 5. The basis for the assignment of the probabilities of the different values is rather weak, because of poor knowledge about the process generating y, that is, because of large epistemic uncertainties. Using precise probabilities, the analyst is required to specify five values, that is, one probability for each of the possible values of y, 1, 2, 3, 4, or 5 (so, actually, only four probabilities are needed as their sum must be 1). Based on the scarce knowledge of the process, the analyst might find it difficult to assign a specific, precise probability mass to each of the five possible values of y. A distribution like, for example, 0.1, 0.2, 0.3, 0.1, 0.3 may leave the analyst with the uneasy feeling that the numbers are somewhat arbitrary given the rather weak background knowledge for the assignments. Yet, the assessor may find it practically feasible, because he or she has just four numbers to assign and feels that the values assigned are indeed reflecting his or her best judgment. Alternatively, the assessor could opt for a uniform distribution to reflect his or her same degree of belief in y being equal to, say, 2 as in y being equal to, say, 5 or any other value. On the premise that such a distribution does, in fact, reflect the assessor's uncertainties, it is attractive to use it, as only one value is needed.

In principle, the assignment of intervals would seem more fitting to a situation of scarce knowledge of the underlying process, which leads to a lack of precision in the value assignment of probabilities. This is because the analyst need not specify one exact number; on the contrary, the analyst is given a way to reflect his or her limited knowledge and associated uncertainty in the assignments, through an imprecise (interval) specification. However, the fact is that in practice the analyst is still required to assign numbers, and not just one for each y but two: a lower bound and an upper bound, reflecting the imprecision of the assignment in view of the scarce knowledge. Take for example $y = 3$. The question to address is: How likely is it that $y = 3$? The analyst will find it difficult to specify just one number in the interval $[0, 1]$, so how, then, does he or she assign the bounds? One way is direct assignment, for instance, the probability of $y = 3$ lies in the interval $[0.2, 0.5]$; but, again, one may feel that the assignment of such an interval is somewhat arbitrary and now, actually, we have a double assignment (lower and upper bounds of the intervals), with a double source of arbitrariness. Also, it is challenging to really interpret what the assigned interval expresses. First of all, it expresses that the analyst is not willing to specify his or her degree of belief on the probability of $y = 3$ more precisely than $[0.2, 0.5]$. It also states that the analyst's degree of belief on the realization of the value $y = 3$ is higher than when drawing a specific ball out of an urn having two balls, but lower than when drawing a ball from an urn having five balls. The analyst is not willing to specify his or her beliefs on the value of y further than this. This reasoning is not straightforward and it might be difficult to "absorb" by assessors who are asked to provide the assignments. In our experience, the assessors need a lot of training and practice to get used to this way of thinking.

In addition, many assessors struggle to understand what is gained by the use of interval probability assignments compared to exact numbers. In the end, the richer description provided by the intervals, which are capable of representing the imprecision in the assignments, is contained in a double set of numbers, whose message is far more difficult to read than single point values. With the difficulties that they encounter in assigning the numbers, assessors question the need for going beyond one uncertainty description level, as provided by single-valued probabilities.

This reflection is important for the practice of risk assessment and the consequential decision making. To be able to effectively use interval probabilities in practice, the obstacles in the interval assignments need to be identified, addressed, and adequately dealt with. More, and extensive, research needs to be carried out in the areas of interval elicitations and interval interpretations. Until solid solutions are offered, pragmatically, our recommendations for cases in which the background knowledge is poor and not given in a clear structured form are as follows:

- Continue to use exact probabilities, but supplemented with an explicit characterization of the background knowledge, for example, by a qualitative approach for assessing the importance of the assumptions that the quantitative analysis is based on (see Section 7.5).

- Use interval probabilities to supplement the exact probabilities if the format of the information and knowledge available justifies the assignment of a specific interval of values, for example, if a specific possibility function can be derived, or the expert judgments are elicited in a form that supports the use of intervals. For instance, suppose that the experts provide information about y of the form: there is nothing to suggest that the value of y is or is not 3; for the remaining y values 1, 2, 4, and 5, the upper probabilities are 0.2, 0.5, 0.7, and 0.3, respectively. These elicited expert judgments could be well reflected by the set of probability intervals $[0, 0.2]$, $[0, 0.5]$, $[0, 1]$, $[0, 0.7]$, and $[0, 0.3]$.

7.3 Non-probabilistic representations with interpretations other than lower and upper probabilities

Imprecise probability analysis can be seen as an extension of probability analysis. Its common starting ground is that single-valued probability is not considered adequate to represent uncertainty and the solution called for is a representation of uncertainty based on measures interpreted as lower and upper probabilities. Also the representations based on evidence theory (belief and plausibility measures) and possibility theory (necessity and possibility measures) can be interpreted as lower and upper probabilities; in fact, technically both possibility theory and probability theory are special cases of evidence theory. However, belief measures and possibility measures can also be understood as expressing "degrees of belief" and "degrees of possibility" per se, and not as lower and upper probabilities. This is how belief functions are to be understood according to Shafer (1976), who presents evidence theory as a generalization of the Bayesian theory of subjective probability in the sense that it does not require probabilities for each proposition or event of interest, but bases the belief in the truth of a proposition or occurrence of an event on the probabilities of other propositions or events related to it. Shafer (1990) uses several metaphors for assigning (and hence interpreting) a belief function Bel. According to the simplest one, $Bel(A) = q$ means that the assessor judges that the strength of the evidence indicating that event A is true is comparable to that of the evidence provided by a witness who has a $q \times 100\%$ chance of being reliable, that is,

$$Bel(A) = P(\text{the witness claiming that } A \text{ is true is reliable}).$$

Hence, it is clear that a duality in terms of interpretation, analogous to that which affects probability (limiting relative frequency vs. degree of belief), also affects possibility theory (degrees of necessity/possibility vs. lower/upper probabilities) and the theory of belief functions (degrees of belief/plausibility vs. lower/upper probabilities). Developing methods based on interpretations other than with reference to lower/upper probabilities represents a distinct development direction for uncertainty representation in risk analysis. Phrases such as "degree of possibility" in possibility theory and "degree of belief" in evidence theory do not provide sufficiently clear interpretations. This is the motivation for Cooke (2004) to ask for an "operational definition" of the possibility function in possibility theory (and also the membership function in fuzzy set theory). One key challenge is therefore to develop clear interpretations (operational definitions) of these concepts, and then to develop appropriate measurement/elicitation procedures.

7.4   Hybrid representations of uncertainty

Probability-bound analysis (mentioned in Section 1.5.2) is an example of how research has also been directed toward the combination of different representations of uncertainty, in this case probabilistic analysis and interval analysis. Another example is probabilistic analysis and possibility theory, where the uncertainties related to some of the parameters of a model are represented by probability distributions and the uncertainties related to the remaining parameters are represented by possibility distributions; see, for example, Baudrit (2006) and the applications in Helton, Johnson, and Oberkampf (2004), Baraldi and Zio (2008), Li and Zio (2012), Pedroni *et al.* (2013), and Flage *et al.* (2013). Probability and possibility distributions are special cases of belief functions, and integrative work has also been carried out in the framework of belief functions theory; see, for example, Dempster and Kong (1988), Almond (1995), and Démotier, Schön, and Denoeux (2006). A hybrid method has also been developed to combine non-parametric predictive inference (NPI) and the standard Bayesian framework; see Montgomery (2009).

The combination of uncertainty representations implies that different representations apply to different situations. Unfortunately, authoritative guidance is hard to find on when to use probability and when to use alternative representations in the context of risk assessment. The argument often seems to be that probability is the appropriate representation of uncertainty only when a sufficient amount of data exists on which to base the assignment of the probability (distribution) in question; however, as mentioned in Chapter 1.5.2, it is not obvious how to make such a prescription operational (Flage, 2010, p. 33). Consider the representation of uncertainty about the parameter(s) of a frequentist probability model. If a sufficiently large amount of data exists, there would be no uncertainty about the parameter(s) and hence no need for a representation of such uncertainty. So when is there enough data to justify probability, but not enough to accurately specify the true value of the parameter in question, and thus make single-valued probability as an epistemic concept superfluous?

As an example (Flage, 2010, p. 33), consider the representation of uncertainty about the parameter p of a Bernoulli random quantity Y_1 for which no observations are available and for which it is considered difficult to have any well-founded opinions. A typical ignorance prior for p would be a beta probability distribution with parameters y_0 and n_0 both equal to 1, which yields a uniform probability distribution on the unit interval and $P(Y_1 = 1) = E[p] = y_0/(y_0 + n_0) = 0.5$. The core of an often-made argument is then that, for the toss of an untested but unsuspected coin, say, most people would also assign $P(Y = 1) = E[q] = 0.5$ (resulting from a non-uniform probability distribution on q, centered around 0.5), where Y equals 1 if the outcome is "heads" and 0 if "tails," but perhaps would have a (qualitatively) vastly different comprehension of the situation. One possible resolution of this problem is to assign, say, $P(Y_1 = 1) = 1/2$ and $P(Y = 1) = 100/200$, with the denominator and numerator chosen to reflect the confidence in the probability assignment (Lindley, 1985); however, this involves the assignment of two numbers, just like the assignment of lower and upper probabilities. Suppose that we observe n realizations, $y_1, \ldots, y_n$, of the Bernoulli process, and through Bayesian updating obtain

$$ p_n = P(Y_{n+1} = 1 \mid y_0, y_1, \ldots, y_n) = \frac{(y_0 + y_1 + \cdots + y_n)}{(y_0 + n_0 + n)}. $$

As n tends to infinity, we have, by the law of large numbers, that p_n tends to the true value of p. The questions, then, are:

- For which values of n is a probabilistic representation justified and for which values not?

- And when a probabilistic representation is not justified, what should be the criteria for selecting a particular representation format (interval analysis, imprecise probability, possibility theory, evidence theory, etc.)?

For the first question a pragmatic approach is probably required. Precise probability is an ideal case where no imprecision is involved; however, there will always be some degree of imprecision. On the other hand, because of the relative simplicity of using (calculating with) probabilities, it is desirable to use probability if the level of imprecision involved is considered sufficiently small. Also, depending on the relevance of the available observations, different values of n could be judged as sufficient in different situations.

The second question points to an important direction of research in relation to the development of hybrid approaches, as well as for specific hybrid methods (interval analysis/probability, possibility/probability, etc.).

7.5 Semi-quantitative approaches

The representations described so far are all quantitative. Another approach, which may be referred to as semi-quantitative, is based on a mixture of quantitative representations and qualitative methods. It may, therefore, be considered a type of hybrid approach, integrating quantitative and qualitative methods. Specifically a semi-quantitative approach consists of using a quantitative uncertainty representation (1–4 above), supplemented by a qualitative assessment and characterization of the background knowledge K of the output, to capture aspects beyond what can be transformed into and expressed in quantitative form.

Examples of approaches of this type are those described in Aven (2008a, 2008b, 2013b) and Flage and Aven (2009): in both examples, standard probabilistic risk descriptions are supplemented by a qualitative assessment of uncertainty aspects not properly reflected by the quantitative descriptions.

A semi-quantitative approach also stands on the belief that probability is not perfect, and more strongly it implies the belief that the full scope of uncertainty and risk cannot be transformed into a quantitative format, using probability or any other measure of uncertainty. On this basis, "uncertainty factors" that are "hidden" in the background knowledge of the subjective probabilities are identified and assessed in a qualitative way. In terms of the notation introduced in Part I, the uncertainty characterization could, for example, be in the format $Q = (P, U_F)$, where U_F denotes a qualitative characterization of uncertainty factors "hidden" in the background knowledge K on which P is conditional.

This identification and assessment are made using simple procedures to categorize the strength of the knowledge that supports the probabilistic analysis. For example, a judgment is made that the background knowledge is weak if one or more of these conditions are true (Flage and Aven, 2009):

- The assumptions made represent strong simplifications.

- Data is not available, or is unreliable.

- There is lack of agreement/consensus among experts.

- The phenomena involved are not well understood; models are non-existent or known/believed to give poor predictions.

If, on the other hand, all of the following conditions are met, the knowledge is considered strong:

- The assumptions made are seen as very reasonable.

- More reliable data is available.

- There is broad agreement/consensus among experts.

- The phenomena involved are well understood; the models used are known to give predictions with the required accuracy.

Cases in between are classified as having medium strength of knowledge. More elaborate schemes can be developed based on judgments according also to the criticality of the assumptions (Aven, 2013b; Flage and Aven, 2009). The idea is to perform a crude risk assessment of potential deviations from the conditions/states defined by the assumptions. The aim of the assessment is to assign a risk score for each deviation, which reflects the risk related to the magnitude of the deviation and its implications. This "assumption–deviation risk" score provides a measure of criticality or importance of the assumption. Depending on an overall judgment of these assumptions scores, a total strength of knowledge level is determined. For further details, see the example below.

This criticality (importance) scoring of assumptions can be used as a guideline for where to place the focus to improve the risk assessment. The assumptions with the high criticality scores should be examined to see if they can be dealt with in some way and removed from the highest importance category (e.g., using the law of total probability). There will, however, always be some factors on which a probability or other risk metrics cannot be made unconditional, for example, data and current phenomenological understanding.

Example 7.1   A risk assessment of a LNG plant (taken from Aven, 2013b)

We return to the example introduced in Chapter 1.5.1.1, where an LNG plant is planned and the operator would like to locate it no more than a few hundred meters from a residential area (Vinnem, 2010). Several quantitative risk assessments (QRAs) are performed in order to demonstrate that the risk is acceptable according to some predefined risk acceptance criteria. In the QRAs risk is expressed using computed probabilities and expected values. The risk metrics used cover both individual risk and f–n curves. It turns out that the assessments and the associated risk management meet strong criticism. The neighbors and many independent experts find the risk characterization insufficient – they argue that risk has been reported according to a too narrow risk perspective.

Risk is described in this case through risk matrices, individual risk (IR) numbers, and f–n curves. These curves show the assigned probability for accidents occurring with a least x fatalities as a function of x (Bedford and

Cooke, 2001). To these curves, we add a dimension reflecting the strength of knowledge on which these assignments are based. The same is the case for the individual risk, which expresses the assessor's probability that an arbitrary but specific person shall be killed during a specific year. As for the f–n curve, we need to add the strength of knowledge component.

For this particular case, the strength of knowledge is assessed by reviewing the list of assumptions and assessing the importance of these according to the procedure outlined above. To compute the risk metrics, a number of assumptions are made, for example (Aven, 2011a):

1. The event tree model.

2. A specific number of exposed people.

3. A specific fraction of fatalities in different scenarios.

4. The probabilities and frequencies of leakages were based on a database for offshore hydrocarbon releases.

5. All vessels and piping are protected by water-based systems (monitors, hydrants).

6. In the event of an impact from a passing vessel on a LNG tanker loading at the quay, the gas release would be ignited immediately (by sparks generated by the collision itself).

Assumption–deviation risk scoring of these (and many other assumptions) (uncertainty factors) were assigned, see Table 7.2 (only the scores of these six assumptions are shown). First an assessment was carried out using the criteria by Flage and Aven (2009) mentioned above this example, then the more detailed method was applied where feasible. Let us look at one of these assumptions, the last one.

Table 7.2   Assumption–deviation risk scoring of six assumptions made in the risk assessment of the LNG plant.

High risk		×	×			×
Moderate risk	×					
Low risk				×	×	
Assumption	1	2	3	4	5	6

Assumption 6, among others, was given the highest risk score. Several experts argued against this assumption. One of these wrote:

> The implication of this assumption was that it was unnecessary to consider in the studies any spreading of the gas cloud due to wind and heating of the liquefied gas, with obvious consequences for the scenarios the public might be exposed to. Such a very critical assumption should at least have been

subjected to a sensitivity study in order to illustrate how changes in the assumption would affect the results, and the robustness of the assumption discussed. None of this, however, has been provided in any of the studies.

(Vinnem, 2010)

An uncertainty factor related to assumption 6 can be formulated as: the time duration between the gas release and the ignition for the relevant scenario. In the assumption–deviation risk assessment, we question how the consequences C are influenced by increasing this duration from zero to (say) one hour, and how likely it is to have such a deviation. A high score was assigned for this assumption, as both the deviation probability and the consequence of the deviation were considered to be quite large. The strength of knowledge judgments of the triplet assignments (magnitude of the deviation, probability of this magnitude to occur, the effect of the change on the consequences) was considered to be moderately large, and this further supported the high risk score for this assumption.

Due to the many assumptions giving a rather high risk/criticality score, the overall conclusion was that the strength of knowledge of the analysis needs to be classified as weak or at best medium. This analysis and conclusion would be essential to accompany the numerical results produced. Obviously, care has to be taken when comparing the produced f–n curves and the individual risk numbers to predefined risk acceptance criteria, without also reflecting on the strong dependencies of the assumptions made.

This risk/criticality scoring can also be used as a guide on where to place the focus to improve the risk assessment. The assumptions with the high score should be examined to see if they can be dealt with in some way and removed from the highest risk/criticality category. However, in practice it is never possible to carry out a quantitative risk assessment without making many assumptions.

Using the general terminology (C', Q, K) introduced in Part I and letting D denote the deviation, the deviation risk can be expressed as $(\Delta C', Q, KD)$, where $\Delta C'$ is the change in the consequences (which includes D) and KD is the knowledge that $\Delta C'$ and Q are based on. In a practical procedure, a score is first based on a judgment of D and related probabilities P (i.e., $Q = P$) (keeping in mind the implications on C'), and adjusted by a consideration of the strength of the knowledge KD.

The issue of uncertainties in the background knowledge is addressed by Mosleh and Bier (1996). They refer to a subjective probability $P(A \mid Y)$ expressing the probability of the event A given a set of conditions Y, and argue that the concept of "uncertainty about probabilities" is meaningful in the sense that $P(A \mid Y)$, seen as a function of Y, is uncertain (it is a random variable). Consequently, there is uncertainty about the random probability $P(A \mid Y)$.

However, it should be noted that the probability is not an unknown quantity (random variable) for the analyst. The probability $P(A \mid K)$ is conditional on the background knowledge K, and some aspects of this knowledge K can be linked to Y as described by Mosleh and Bier (1996). The analyst has decided to assign his or her probability based on K. If the analyst finds that the uncertainty about Y should be taken into account, he or she would modify the assigned probability using the law of total probability. From this it does not follow, however, that $P(A \mid K)$ is uncertain, as such a statement would presume that a true probability exists. The assessor needs to make clear what is uncertain and subject to the uncertainty assessment and what forms the background knowledge. One may think that it is possible and desirable to remove all such Ys from K, but it is impossible in a practical risk assessment context. There is always a need for some type of background knowledge to base our probabilities on, and in many cases this knowledge would not be possible to specify using quantities like Y (Aven, 2011c).

Say that an assumption $Y = y_0$ is made in the risk assessment. Then the assumption–deviation risk referred to above is risk related to the deviation $Y - y_0$. To assess/describe this risk, the crude approach presented above can be used, highlighting the magnitude of the deviation, the (subjective) probability of the magnitude to occur, and the effect of the change on the consequences C addressed in the risk assessment, in line with the standard triplet way of describing risk (Kaplan and Garrick, 1981). In addition an overall judgment of the strength of the background knowledge of this triplet risk assessment is to be made.

References – Part II

Almond, R.G. (1995) *Graphical Belief Models*, Chapman & Hall, London.

Anoop, M.B. and Rao, K.B. (2008) Determination of bounds on failure probability in the presence of hybrid uncertainties. *Sadhana*, **33**, 753–765.

Aven, T. (2008a) *Risk Analysis: Assessing Uncertainties Beyond Expected Values and Probabilities*, John Wiley & Sons, Ltd, Chichester.

Aven, T. (2008b) A semi-quantitative approach to risk analysis, as an alternative to QRAs. *Reliability Engineering and System Safety*, **93**, 768–775.

Aven, T. (2011a) *Quantitative Risk Assessment: The Scientific Platform*, Cambridge University Press, Cambridge.

Aven, T. (2011b) Selective critique of risk assessments with recommendations for improving methodology and practice. *Reliability Engineering and System Safety*, **96**, 509–514.

Aven, T. (2011c) Interpretations of alternative uncertainty representations in a reliability and risk analysis context. *Reliability Engineering and System Safety*, **96**, 353–360.

Aven, T. (2012a) *Foundations of Risk Analysis*, 2nd edn, John Wiley & Sons, Ltd, Chichester.

Aven, T. (2013a) How to define and interpret a probability in a risk and safety setting. *Safety Science*, **51**, 223–231. Discussion paper, with general introduction by Associate Editor Genserik Reniers.

Aven, T. (2013b) Practical implications of the new risk perspectives. *Reliability Engineering and System Safety*, **115**, 136–145.

Aven, T. and Zio, E. (2011) Some considerations on the treatment of uncertainties in risk assessment for practical decision-making. *Reliability Engineering and System Safety*, **96** (1), 64–74.

Baraldi, P. and Zio, E. (2008) A combined Monte Carlo and possibilistic approach to uncertainty propagation in event tree analysis. *Risk Analysis*, **28**, 1309–1326.

Baraldi, P., Pedroni, N., Zio, E. *et al.* (2011) Monte Carlo and fuzzy interval propagation of hybrid uncertainties on a risk model for the design of a flood protection dike. Proceedings of European Safety and Reliability Conference (ESREL 2011), Troyes, France, 18–22 September, pp. 2167–2175.

Baudrit, C. and Dubois, D. (2006) Practical representations of incomplete probabilistic knowledge. *Computational Statistics & Data Analysis*, **51** (1), 86–108.

Baudrit, C., Dubois, D., and Guyonnet, D. (2006) Joint propagation and exploitation of probabilistic and possibilistic information in risk assessment. *IEEE Transactions on Fuzzy Systems*, **14** (5), 593–608.

Bedford, T. and Cooke, RM. (2001) *Probabilistic Risk Analysis: Foundations and Methods*, Cambridge University Press, Cambridge.

Bernardo, J.M. and Smith, A.F.M. (1994) *Bayesian Theory*, John Wiley & Sons, Ltd, Chichester.

Boole, G. (1854) *An Investigation of the Laws of Thought on Which are Founded the Mathematical Theories of Logic and Probabilities*, Walton and Maberly, London, http://www.gutenberg.org/etext/15114.

Carnap, R. (1922) *Der logische Aufbau der Welt*, Berlin.

Carnap, R. (1929) *Abriss der Logistik*, Wien.

Cheney, E.W. (1966) *Introduction to Approximation Theory*, McGraw-Hill, New York.

Cooke, R.M. (1986) Conceptual fallacies in subjective probability. *Topoi*, **5**, 21–27.

Cooke, R.M. (2004) The anatomy of the squizzel: the role of operational definitions in representing uncertainty. *Reliability Engineering and System Safety*, **85**, 313–319.

Coolen, F.P.A. (2004) On the use of imprecise probabilities in reliability. *Quality and Reliability Engineering International*, **20**, 193–202.

Coolen, F.P.A. and Utkin, L.V. (2007) Imprecise probability: a concise overview, in *Risk, Reliability and Societal Safety: Proceedings of the European Safety and Reliability Conference (ESREL), Stavanger, Norway, 25–27 June 2007* (eds. T. Aven and J.E. Vinnem), Taylor & Francis, London, pp. 1959–1965.

Coolen, F.P.A., Troffaes, M.C.M., and Augustin, T. (2010) Imprecise probability, in *International Encyclopedia of Statistical Science*, Springer Verlag, Berlin.

Couso, I., Moral, S., and Walley, P. (1999) Examples of independence for imprecise probabilities. Proceedings of the 1st International Symposium on Imprecise Probabilities and Their Applications (ISIPTA 1999), University of Ghent, Belgium (eds. G.De Cooman, F.G. Cozman, S. Moral, and P. Walley), pp. 121–130.

Cullen, A.C. and Frey, H.C. (1999) *Probabilistic Techniques in Exposure Assessment: A Handbook for Dealing with Variability and Uncertainty in Models and Inputs*, Plenum Press, New York.

Cowell, R.G., Dawid, A.P., Lauritzen, S.L., and Spiegelhalter, D.J. (1999) *Probabilistic Networks and Expert Systems*, Springer Verlag, New York.

de Finetti, B. (1930) Fondamenti logici del ragionamento probabilistico. *Bollettino dell'Unione Matematica Italiana*, **5**, 1–3.

de Finetti, B. (1974) *Theory of Probability*, John Wiley & Sons, Inc., New York.

de Laplace, P.S. (1812) *Théorie analytique des probabilités*, Courcier Imprimeur, Paris.

Démotier, S., Schön, W., and Denoeux, T. (2006) Risk assessment based on weak information using belief functions: a case study in water treatment. *IEEE Transactions on Systems, Man, and Cybernetics*, **36**, 382–396.

Dempster, A.P. (1967) Upper and lower probabilities induced by a multivalued mapping. *Annals of Mathematical Statistics*, **38**, 325–339.

Dempster, A.P. and Kong, A. (1988) Uncertain evidence and artificial analysis. *Journal of Statistical Planning and Inference*, **20** (3), 355–368.

Dubois, D. (2006) Possibility theory and statistical reasoning. *Computational Statistics & Data Analysis*, **51**, 47–69.

Dubois, D. (2010) Representation, propagation and decision issues in risk analysis under incomplete probabilistic information. *Risk Analysis*, **30**, 361–368.

Dubois, D. and Prade, H. (1988) *Possibility Theory*, Plenum Press, New York.

Dubois, D. and Prade, H. (2007) Possibility theory. *Scholarpedia*, **2** (10), 2074.

Dubois, D., Nguyen, H.T., and Prade, H. (2000) Fuzzy sets and probability: misunderstandings, bridges and gaps, in *Fundamentals of Fuzzy Sets* (eds. D. Dubois and H. Prade), Kluwer Academic, Boston, MA, pp. 343–438.

Dubucs, J.-P. (1993) *Philosophy of Probability*, Kluwer Academic, Dordrecht.

Ferson, S. and Ginzburg, L.R. (1996) Different methods are needed to propagate ignorance and variability. *Reliability Engineering and System Safety*, **54**, 133–144.

Flage, R. (2010) Contributions to the treatment of uncertainty in risk assessment and management. PhD thesis No. 100. University of Stavanger.

Flage, R. and Aven, T. (2009) Expressing and communicating uncertainty in relation to quantitative risk analysis (QRA). *Reliability & Risk Analysis: Theory & Applications*, **2** (13), 9–18.

Flage, R., Baraldi, P., Ameruso, F. *et al.* (2009) Handling epistemic uncertainties in fault tree analysis by probabilistic and possibilistic approaches. European Safety and Reliability Conference (ESREL 2009), Prague, Czech Republic, 7–10 September 2009, pp. 1761–1768.

Flage, R., Baraldi, P., Aven, T., and Zio, E. (2013) Probabilistic and possibilistic treatment of epistemic uncertainties in fault tree analysis. *Risk Analysis*, **33** (1), 121–133.

Flage, R., Aven, T., Zio, E., and Baraldi, P. (submitted) Concerns, challenges and directions of development for the issue of representing uncertainty in risk assessment. *Risk Analysis*.

Franklin, J. (2001) Resurrecting logical probability. *Erkenntnis*, **55** (2), 277–305.

Gaines, B.R. and Kohout, L. (1975) Possible automata. Proceedings of the International Symposium on Multiple-Valued Logics, Bloomington, IN, USA, pp. 183–196.

Gillies, D. (2000) *Philosophical Theories of Probability*, Routledge, London.

Hajek, A. (2001) Probability, logic and probability logic, in *The Blackwell Companion to Logic* (ed. Lou Goble), pp. 362–384.

Helton, J.C., Johnson, J.D., and Oberkampf, W.L. (2004) An exploration of alternative approaches to the representation of uncertainty in model predictions. *Reliability Engineering and System Safety*, **85** (2), 39–71.

Kaplan, S. and Garrick, B.J. (1981) On the quantitative definition of risk. *Risk Analysis*, **1** (1), 11–27.

Keynes, J. (1921) *Treatise on Probability*, London.

Kiureghian, D., Lin, H.Z., and Hwang, S.J. (1987) Second-order reliability approximations. *Journal of Engineering Mechanics*, **113** (8), 1208–1225.

Klir, G.J. (1998) *Uncertainty-Based Information: Elements of Generalized Information Theory*, Springer Verlag, Heidelberg.

Kolmogorov, A.N. (1933) *Grundbegriffe der Wahrscheinlichkeitrechnung*, Ergebnisse Der Mathematik, trans. as *Foundations of Probability*, Chelsea, New York, 1950.

Kozine, I. and Utkin, L. (2002) Processing unreliable judgements with an imprecise hierarchical model. *Risk Decision and Policy*, **7**, 1–15.

Kuznetsov, V.P. (1991) *Interval Statistical Models* (in Russian), Radio i Svyaz, Moscow.

Li, Y. and Zio, E. (2012) Uncertainty analysis of the adequacy assessment model of a distributed generation system. *Renewable Energy*, **41**, 235–244.

Limbourg, P. and de Rocquigny, E. (2010) Uncertainty analysis using evidence theory – confronting level-1 and level-2 approaches with data availability and computational constraints. *Reliability Engineering and System Safety*, **95** (5), 550–564.

Lindley, D.V. (1985) *Making Decisions*, 2nd edn, John Wiley & Sons, Ltd, London.

Lindley, D.V. (2000) The philosophy of statistics. *The Statistician*, **49** (3), 293–337.

Lindley, D.V. (2006) *Understanding Uncertainty*, John Wiley & Sons, Inc., Hoboken, NJ.

Montgomery, V. (2009) New statistical methods in risk assessment by probability bounds. PhD thesis. Durham University.

Morgan, M.G. and Henrion, M. (1990) *Uncertainty: A Guide to Dealing with Uncertainty in Quantitative Risk and Policy Analysis*, Cambridge University Press, Cambridge.

Mosleh, A. and Bier, V. (1996) Uncertainty about probability: a reconciliation with the subjectivist viewpoint. *IEEE Transactions on Systems, Man, and Cybernetics. Part A – Systems and Humans*, **26** (3), 303–310.

North, D.W. (2010) Probability theory and consistent reasoning. *Risk Analysis*, **30** (3), 377–380.

Nuclear Regulatory Commission (NRC) (2009) NUREG-1855 – Guidance on the Treatment of Uncertainties Associated with PRAs in Risk-Informed Decision Making. Main Report. Vol. 1.

O'Hagan, A. and Oakley, J.E. (2004) Probability is perfect, but we can't elicit it perfectly. *Reliability Engineering and System Safety*, **85**, 239–248.

Pedroni, N. and Zio, E. (2012) Empirical comparison of methods for the hierarchical propagation of hybrid uncertainty in risk assessment, in presence of dependences. *International Journal of Uncertainty, Fuzziness and Knowledge-Based Systems*, **20** (4), 509–557.

Pedroni, N., Zio, E., Ferrario, E. *et al.* (2013) Hierarchical propagation of probabilistic and non-probabilistic uncertainty in the parameters of a risk model. *Computers & Structures*, in press.

Ramsey, F. (1931) Truth and probability, in *Foundations of Mathematics and Other Logical Essays*, London, Routledge & Kegan Paul.

Scheerlinck, K., Vernieuwe, H., and De Baets, B. (2012) Zadeh's extension principle for continuous functions of non-interactive variables: a parallel optimization approach. *IEEE Transactions on Fuzzy Systems*, **20** (1), 96–108.

Shafer, G.A. (1976) *Mathematical Theory of Evidence*, Princeton University Press, Princeton, NJ.

Shafer, G.A. (1990) Perspectives on the theory and practice of belief functions. *International Journal of Approximate Reasoning*, **4**, 323–362.

Singpurwalla, N.D. (2006) *Reliability and Risk: A Bayesian Perspective*, John Wiley & Sons, Ltd, Chichester.

Singpurwalla, N.D. and Wilson, A.G. (2008) Probability, chance and the probability of chance. *IIE Transactions*, **41**, 12–22.

Smets, P. (1994) What is Dempster–Shafer's model? in *Advances in the Dempster–Shafer Theory of Evidence* (eds. R.R. Yager, M. Fedrizzi, and J. Kacprzyk), John Wiley & Sons, Inc., San Mateo, CA, pp. 5–34.

Springer, M.D. (1979) *The Algebra of Random Variables*, John Wiley & Sons, Inc., New York.

Stanford Encyclopedia of Philosophy (SEP) (2009) Interpretations of probability, http://plato .stanford.edu/entries/probability-interpret/ (accessed May 22, 2011).

Su, Z.G., Wang, P.G., Yu, X.J., and Lv, Z.Z. (2011) Maximal confidence intervals of the interval-valued belief structure and applications. *Information Sciences*, **181** (9), 1700–1721.

van Lambalgen, M. (1990) The axiomatisation of randomness. *Journal of Symbolic Logic*, **55** (3), 1143–1167.

Vinnem, J.E. (2010) Risk analysis and risk acceptance criteria in the planning processes of hazardous facilities – a case of an LNG plant in an urban area. *Reliability Engineering and System Safety*, **95** (6), 662–670.

Walley, P. (1991) *Statistical Reasoning with Imprecise Probabilities*, Chapman & Hall, London.

Weichselberger, K. (2000) The theory of interval-probability as a unifying concept for uncertainty. *International Journal of Approximate Reasoning*, **24**, 149–170.

Yager, R.R. (2011) On the fusion of imprecise uncertainty measures using belief structures. *Information Sciences*, **181**, 3199–3209.

Xiong, F., Greene, S., Chen, W., Xiong, Y., and Yang, S. (2010) A new sparse grid based method for uncertainty propagation. *Structural and Multidisciplinary Optimization*, **41** (3), 335–349.

Zadeh, L.A. (1975) The concept of a linguistic variable and its application to approximate reasoning. *Information Science*, **8**, 199–249.

Zadeh, L.A. (1978) Fuzzy sets as a basis for a theory of possibility. *Fuzzy Sets and Systems*, **1**, 3–28.

Zio, E. (2007) *An Introduction to the Basics of Reliability and Risk Analysis*, World Scientific, Hackensack, NJ.

Zio, E. (2013) *The Monte Carlo Simulation Method for System Reliability and Risk Analysis, Springer Series in Reliability Engineering*, Springer Verlag, Berlin.

Part III

PRACTICAL APPLICATIONS

In this third part of the book, we present some practical applications of the methods introduced in Part II for the representation and propagation of uncertainty in risk assessment. The applications range from the reliability, availability, and maintainability (RAM) analysis of single components to system risk assessment in terms of characterizing uncertainty related to frequentist probabilities of occurrence of undesirable events and of their consequences.

Specifically, Chapters 8 and 9 present applications regarding quantification of the reliability of a structure and the availability of a maintained industrial item, respectively. Chapter 10 focuses on the event tree analysis of a complex multi-component engineering system and Chapter 11 presents an analysis of the consequences of undesirable events associated with the operation of an industrial system. Finally, Chapter 12 concerns the probabilitistic risk assessment of a process plant.

All the applications presented involve assessments in circumstances characterized by poor knowledge of one or more uncertain quantities. In the applications of Chapters 8–11, the assessments are first performed using a non-probabilistic method of uncertainty representation and propagation. Then, the results are transformed into probability distributions for comparison to the outcomes of a purely probabilistic uncertainty propagation treatment. The application in Chapter 12 uses a fully Bayesian approach to represent and propagate uncertainty.

Table III.1 gives a snapshot of the applications presented and the methods of uncertainty representation and propagation used in the different cases.

Uncertainty in Risk Assessment: The Representation and Treatment of Uncertainties by Probabilistic and Non-Probabilistic Methods, First Edition. Terje Aven, Piero Baraldi, Roger Flage and Enrico Zio.
© 2014 John Wiley & Sons, Ltd. Published 2014 by John Wiley & Sons, Ltd.

Table III.1 Structure of Part III.

Chapter	Application	Uncertainty representation		Uncertainty propagation	
		Type of uncertainty	Representation method	Type	Method
8	Structural reliability analysis	Epistemic and aleatory	Frequentist probabilities and possibility distributions	Level 1	Hybrid probabilistic–possibilistic
9	Maintenance performance assessment	Epistemic and aleatory	Frequentist probabilities and evidence theory	Level 2	Hybrid probabilistic–evidence theory
10	Event tree analysis	Epistemic and aleatory	Frequentist probabilities and possibility distributions	Level 2	Hybrid probabilistic–possibilistic
11	Consequence evaluation	Epistemic and aleatory	Frequentist probabilities and possibility distributions	Level 1	Hybrid probabilistic–possibilistic
12	Probabilistic risk assessment	Epistemic and aleatory	Bayesian representation	Level 2	Probabilistic (Bayesian representation)

8

Uncertainty representation and propagation in structural reliability analysis

In this chapter, we consider the problem of uncertainty representation and propagation in the quantification of the reliability of a structure subject to cracking due to fatigue. The application is partly based on Baraldi, Popescu, and Zio (2010b).

8.1 Structural reliability analysis

The reliability of a structure is an important issue which needs to be given due attention during all phases of the life cycle of the structure, including design, manufacturing, installation, operation, maintenance, and demolition. For this, mathematical models are developed to describe the evolution of the physical state of the structure. In particular, degradation models are developed to predict the future degradation evolution of an aging structure. These models must describe the stochastic variation of the physical state of the structure under a changing environment (loads, material properties, etc.), while giving due account to the epistemic uncertainty on the model parameters introduced. In this setting, structural reliability analysis requires representing these uncertainties and propagating them to assess the structure failure probability, typically the probability that the structure's physical state exceeds a specified level which defines the failure criterion (Ardillon, 2010).

8.1.1 A model of crack propagation under cyclic fatigue

Crack propagation under cyclic fatigue is a structural degradation process which arises in several types of materials and involves very different engineering

Uncertainty in Risk Assessment: The Representation and Treatment of Uncertainties by Probabilistic and Non-Probabilistic Methods, First Edition. Terje Aven, Piero Baraldi, Roger Flage and Enrico Zio.
© 2014 John Wiley & Sons, Ltd. Published 2014 by John Wiley & Sons, Ltd.

applications. In this section, we consider pressurized components of nuclear power plant (NPP) systems (Jeong *et al.*, 2005) which are subject to crack propagation initiating from manufacturing flaws under cyclic loading caused by allowed power variations or incidental transients. Once initiated, the degradation process can propagate under stressful operating conditions, up to limits threatening the structural integrity (Mustapa and Tamin, 2004). Other examples of materials which can encounter this kind of degradation process are silicon nitride ceramics used for automobile turbocharger wheels and engine valves, and pyrolytic carbon used for prosthetic cardiac devices (Ritchie, 1999).

We refer to the commonly used Paris–Erdogan model for describing the evolution of a crack in a structure subject to cyclic fatigue. Let h indicate the crack depth and L the load cycle (Pulkkinen, 1991). Then, the following equation describes the evolution of the crack depth as a function of the load cycles:

$$\frac{dh}{dL} = C(\Delta K)^{\eta}, \tag{8.1}$$

where C and η are constants related to the material properties (Kozin and Bogdanoff, 1989; Provan, 1987), which can be estimated from experimental data (Bigerelle and Iost, 1999), and ΔK is the stress intensity amplitude, roughly proportional to the square root of h (Provan, 1987):

$$\Delta K = \beta\sqrt{h}, \tag{8.2}$$

where β is again a constant which may be determined from experimental data. According to Provan (1987), the intrinsic stochasticity of the process is introduced into the model by modifying (8.1) as follows:

$$\frac{dh}{dL} = e^{\omega}C\left(\beta\sqrt{h}\right)^{\eta}, \tag{8.3}$$

where $\omega \sim N(0, \sigma_{\omega}^2)$ is a normally distributed random variable with zero mean and variance σ_{ω}. For ΔL sufficiently small, this model can be discretized to give (Pulkkinen, 1991):

$$h_k = h_{k-1} + e^{\omega_k}C(\Delta K)^{\eta}\Delta L \tag{8.4}$$

which represents a nonlinear Markov process with independent, non-stationary degradation increments.

8.2 Case study

We consider the problem of assessing the reliability of a structure subject to the propagation of a crack due to fatigue. We consider the model in (8.3) to describe the evolution of the crack depth and we assume that a load cycle is applied every time step.

Table 8.1 Numerical values of the case study parameters.

Parameter	Value
ω_k	A random variable with distribution $N(0, \ 1.7)$
C	0.005
β	1
η	1.3
T_{miss}	500 time units
H_{max}	An uncertain variable whose value lies in the range $\in [9, \ 11]$, with most likely value 10

The structure is designed to accomplish its function for a certain time window, called mission time T_{miss}, and it is considered as failed when the value of the degradation exceeds a certain threshold, denoted H_{max}. We assume that the degradation threshold H_{max} is subject to epistemic uncertainty due to limited information on the failure mechanism, but according to an expert H_{max} lies in the range of $[9, \ 11]$ with a most likely value of 10 (in arbitrary units).

The reliability estimations are performed assuming that at the present time t_p an exact measurement of the current degradation level is available and is given by the crack depth $h(t_p)$. The results that follow refer to the initial condition $h(0) = 1$.

Table 8.1 presents values of the parameters defining the degradation process considered in this application.

An uncertain Boolean variable X_S is introduced to indicate the state of the structure at mission time: 0 indicates failure and 1 success (Baraldi, Popescu, and Zio, 2010b). The variable X_S depends on the crack depth at mission time, that is, $h(T_{\text{miss}})$, which is subject to aleatory uncertainty, and on the failure threshold H_{max}, which is subject to epistemic uncertainty. Hence, X_S is a function g of h and H_{max}, as given in (8.5):

$$X_S = g(h(T_{\text{miss}}), H_{\text{max}}) = \begin{cases} 1 & \text{if } h(T_{\text{miss}}) < H_{\text{max}} \\ 0 & \text{otherwise.} \end{cases} \tag{8.5}$$

It is assumed that no repairs are possible to the structure. The objective is, then, to assess uncertainty regarding the structure's reliability, given by the frequency probability $P_f(X_S = 1)$ that the structure has not failed prior to mission time.

Application 8.1   (in a nutshell)

Input uncertain quantities:

- crack depth of the structure at mission time (it depends on the intrinsically stochastic evolution of the degradation process), $h(T_{\text{miss}})$
- failure threshold, H_{max}.

Output quantity:

- state of the structure at mission time, X_S (1 corresponding to success and 0 to failure).

Model:

$$X_S = g(h(T_{miss}), H_{max}) = \begin{cases} 1 & \text{if } h(T_{miss}) < H_{max} \\ 0 & \text{otherwise.} \end{cases}$$

The assessment regards computation of the structure's reliability, that is, $P_f(X_S = 1)$.

Type of uncertainty on the input quantities:

- aleatory on $h(T_{miss})$
- epistemic on H_{max}.

Uncertainty propagation setting:

- level 1 (the aleatory quantity $h(T_{miss})$ is not subject to any form of epistemic uncertainty).

8.3   Uncertainty representation

In order to represent the epistemic uncertainty associated with the value of the threshold H_{max}, a possibility distribution is used. According to the available information, $\pi(H_{max})$ is set equal to 0 outside the interval $[9, 11]$ since the expert considers these values impossible, whereas $\pi(10)$ is set to 1 since, according to the expert, 10 is the most likely value of H_{max}.

Among the possible distributions that may be used to represent the available uncertain information on H_{max}, the triangular possibility distribution π in Figure 8.1(a) (line with circles) is chosen following the observations reported in Section 3.2.2. Figure 8.1(b) shows the corresponding behavior of the necessity and possibility measures of the interval $(-\infty, x]$, for $9 \leq x \leq 11$, which, according to Section 4.1, can be interpreted as the lower and upper cumulative distributions for the threshold H_{max}. In practice, according to this interpretation, the triangular possibility distribution π induces the family of all the probability distributions with range $[9, 11]$ and mode 10. Some examples of the infinite number of probability distributions included in the family are (see Figure 8.1):

- the uniform distribution with range $[9, 11]$;
- the triangular distribution with range $[9, 11]$ and mode 10, which, in this case, completely overlaps the possibility distribution;
- the trapezoidal distribution with range $[9.5, 10.5]$ and lower and upper modes $[9.7, 10.3]$.

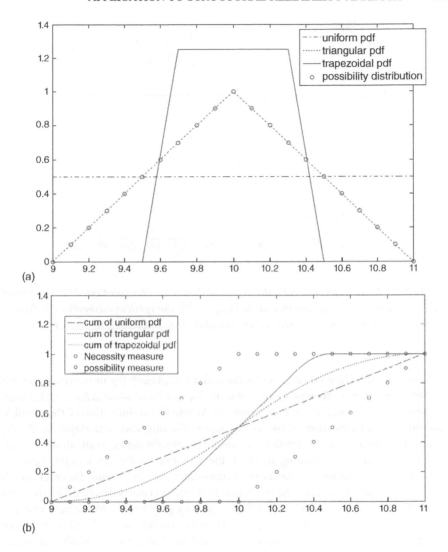

Figure 8.1   (a) Line with circles: possibility distribution representing the uncertainty on H_{max}*; solid line: three examples of probability density functions represented by the possibility distribution. (b) Line with circles: necessity and possibility measures of the set* $(-\infty, x]$ *corresponding to the possibility distribution in (a); solid line: cumulative distributions corresponding to the probability distributions in (a).*

8.4 Uncertainty propagation

The hybrid probabilistic–possibilistic method described and illustrated in Section 6.1.3 is used to propagate the uncertainty from the input quantities: the crack depth at mission time, $h(T_{\text{miss}})$, and the failure threshold, H_{max}, to the structure state at

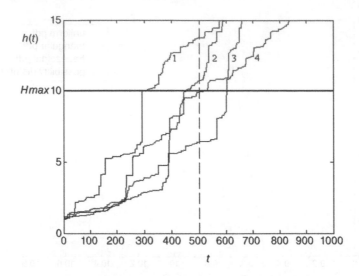

Figure 8.2   Four different realizations of the degradation evolution. The horizontal solid line indicates the failure threshold $H_{max} = 10$; the vertical dashed line indicates the structure's mission time. (Based on Baraldi, Zio, and Popescu, 2008.)

mission time, X_S. The method provides the belief and plausibility measures of the two possible states of the structure at mission time: $X_S = 1$ (success) and $X_S = 0$ (failure).

The application of the methods requires a Monte Carlo simulation of the possible random paths of evolution of the crack depth (Baraldi, Zio, and Popescu, 2008). Figure 8.2 shows four examples of Monte Carlo-simulated realizations of the degradation process. According to (8.4), the variation of the crack depth between two successive time instants tends to increase as time passes and the crack depth increases. Furthermore, since the stochasticity of the process is described by the term e^{ω}, with $\omega \sim N(0, \sigma_{\omega}^2)$, which multiplies a quantity proportional to the current crack depth at time $t, h(t)$, the trajectories tend to be similar at the beginning of the simulation, when the crack depth is small, and tend to have remarkably different evolutions for high values of crack depth.

Note that the uncertainty propagation method is not performing an analytical propagation of the uncertainty (i.e., the variation), but just providing an estimate of the uncertainty (variation) of the output quantity. This estimate is subject to an error that depends on the number, M, of Monte Carlo sampling of the aleatory quantities and the number of α-cuts of the possibility distributions used. In this application, we have set $M = 10^4$ random paths of crack propagation and considered 21 α-cuts of the quantity H_{max}. This choice has been motivated by a trade-off between the accuracy of the estimate and the computational burden: higher values of M and of the number of α-cuts yield a higher accuracy of the output uncertainty distribution estimate, but this causes an increase in the computational burden which is linear with M and the number of α-cuts.

8.5 Results

From the propagation of the uncertainty through the model of (8.3), it turns out that $Bel(\text{success}) = 0.896$ and $Pl(\text{success}) = 0.927$. This result merges the aleatory uncertainty represented by the different realizations of the stochastic degradation process with the epistemic uncertainty represented by the possibility distribution of $H_{\max}$. In line with the interpretation of the belief and plausibility measures as upper and lower probability values, from the assessment performed we have found the interval probability [0.896, 0.927] for the event that the non-repairable structure survives the degradation process up to mission time. The available information does not allow for a more precise value assignment of this probability. Thus, the decision maker is required to cope with this imprecise knowledge, when taking decisions.

8.6 Comparison to a purely probabilistic method

For applying a purely probabilistic approach, the transformation described in Appendix B is applied to the triangular possibility distribution considered in the previous section (Figure 8.1(a), line with circles) to derive the (subjective) probability distribution describing the epistemic uncertainty on $H_{\max}$. The resulting probability density function is

$$f(H_{\max}) = -\frac{1}{2}\log(1 - \pi(H_{\max})). \tag{8.6}$$

Note that, as expected, the maximum value of the probability density function is obtained for $H_{\max} = 10$, which corresponds to the value of $H_{\max}$ for which the possibility distribution is 1. Furthermore, the preference preservation principle according to which

$$\pi(x) > \pi\ (\underline{x}) \Leftrightarrow f(x) > f(\underline{x}) \tag{8.7}$$

is satisfied. The infinite value of the probability density function in $H_{\max} = 10$ is due to the 0 length of the 1 α-cut of $\pi(H_{\max})$, which appears in the denominator of (B.1) (Appendix B).

Figures 8.3(a) and (b) (solid lines) show the obtained probability density function and cumulative distribution, respectively.

From this figure, it can be seen that while the possibilistic representation of the uncertainty on $H_{\max}$ allows us to consider all the possible cumulative distributions lying between the belief and plausibility measures of the set $A = [0, u)$, the probabilistic representation forces us to consider only one specific cumulative distribution, conservatively chosen as the one containing as much uncertainty as possible, in the sense of the highest entropy (due to the possibility–probability transformation). In this sense, the probabilistic density function tends to force information into the representation.

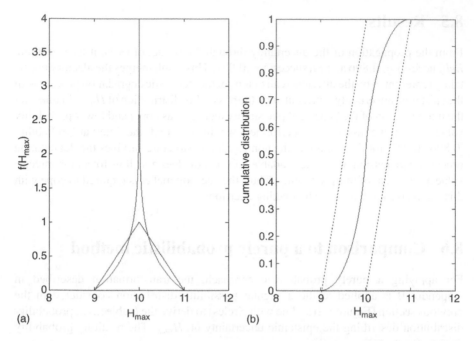

Figure 8.3   (a) Possibilistic distribution of H_{max} (dotted line) and its probabilistic transformation (solid line); (b) corresponding necessity and possibility measures of the set $(-\infty, x]$ (dotted line) and cumulative distribution of H_{max} considered by the purely probabilistic method (solid line). (Based on Baraldi, Zio, and Popescu, 2008.)

The propagation of the uncertainty has, then, been performed according to the double loop method described in Section 5.2.1, which separates the epistemic uncertainty in the outer loop and the aleatory uncertainty in the inner loop. Each time the outer loop is repeated with a value of H_{max} sampled from its probability distribution (Figure 8.3), an estimate of the system reliability is obtained. Thus, upon performing $M_a = 10^4$ simulations of the degradation process for each of the $M_e = 10^3$ values of H_{max}, we obtain 10^3 values of the structure's reliability. These values are used to build the cumulative distributions of the structure's reliability (Figure 8.4). The numbers M_a and M_e have been chosen in order to obtain a computational effort similar to that obtained in the hybrid probability–possibility framework.

From the results, it is possible to choose a level of credibility $(1 - \beta)$ and then provide the credibility interval given by the percentiles $(\beta/2)$ and $(1 - \beta/2)$, such that the subjective probability that the structure's reliability is in the interval is $(1 - \beta)$. For example, for a credibility of $(1 - \beta) = 0.95$, the resulting credibility interval for the structure's reliability is $[0.889, 0.935]$.

In the end, the probabilistic and hybrid methods in the application are shown to lead to similar results for the values of reliability of the structure. However, the interpretation of the results obtained by the probabilistic method requires a degree of

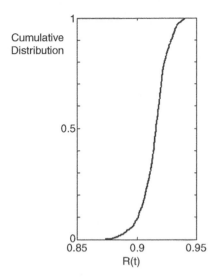

Figure 8.4   Cumulative distribution of the structure's reliability obtained by the purely probabilistic method (based on Baraldi, Zio, and Popescu, 2008).

credibility directly connected to the epistemic uncertainty on the parameters, whereas the hybrid probabilistic–possibilistic method provides information on the limiting cumulative distributions without requiring the definition of credibility levels, giving in this way a more synthetic, albeit less informative, representation of the uncertainty in the structure's reliability.

9

Uncertainty representation and propagation in maintenance performance assessment

In this chapter, we consider the problem of uncertainty representation and propagation in the context of maintenance performance assessment. The application is in part based on Baraldi, Compare, and Zio (2012) and Baraldi, Compare, and Zio (2013a).

9.1 Maintenance performance assessment

In the last few decades, the relevance of maintenance in several sectors of industry has increased in influence on both productivity and safety. For example, maintenance represents a major portion of the total production cost of non-fossil-fuel energy production plants (nuclear, solar, wind, etc.) and its optimization is fundamental for the economic competitiveness of such plants (Zio and Compare, 2013).

In general, the goal of effective maintenance planning is the optimization of production and plant availability, in a way that guarantees safety and respects the associated regulatory requirements (Zio, 2009). For this, several approaches to maintenance modeling, optimization, and management have been proposed in the literature. Usually, these approaches are divided into two main groups: corrective maintenance and scheduled maintenance.

Under the corrective maintenance strategy, the components are operated until failure; then, repair or renovation actions are performed. This is the oldest approach to maintenance and is nowadays still adopted in some industries, especially for equipment which is neither safety critical nor crucial for the production performance of the plant, and whose spare parts are readily available and not expensive (Zio and

Uncertainty in Risk Assessment: The Representation and Treatment of Uncertainties by Probabilistic and Non-Probabilistic Methods, First Edition. Terje Aven, Piero Baraldi, Roger Flage and Enrico Zio.
© 2014 John Wiley & Sons, Ltd. Published 2014 by John Wiley & Sons, Ltd.

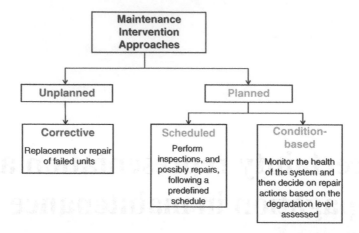

Figure 9.1   Maintenance intervention approaches (based on Baraldi, Zio, and Popescu, 2008).

Compare, 2011). Under some strategies, maintenance is performed on a scheduled basis, which can be predefined (periodic maintenance) or based on information on the degradation state of the equipment (condition-based maintenance). Figure 9.1 shows a sketch of the different maintenance approaches.

In practice, the definition of a proper maintenance plan requires the following (Zio, 2009):

1. The representation of the dynamic interactions among the different components of the system which affect the system behavior and maintenance (e.g., by Petri nets or Bayesian belief networks (Zille *et al.*, 2007)).

2. The proper reliability, maintenance, production, and economic modeling of the involved processes (e.g., by Petri nets, Markov chains, and Monte Carlo simulation (Châtelet, Berenguer, and Jellouli, 2002)).

3. An efficient engine for searching for the potentially optimal maintenance strategies (e.g., by the growing evolutionary computational methods such as genetic algorithms (Marseguerra and Zio, 2000a, 2000b; Marseguerra, Zio, and Martorell, 2006).

4. A solid decision-making-theory structure for their evaluation (Saferelnet, 2006).

In this section, we focus on issues 2 and 3, taking into account the fact that the models developed for this aim typically rely on a number of parameters which are often weakly known in real applications. This is mainly due to the lack of real/field data collected during operation or properly designed tests. In these cases, the main source of information for estimating these parameters is experts' judgment.

9.2   Case study

We consider the definition of a maintenance plan for a check valve of a turbo pump lubricating system in a nuclear power plant (Zille *et al.*, 2009). This component undergoes one principal degradation mechanism, namely, fatigue, and only one failure mode, namely, rupture. The degradation process is modeled as a discrete-state, continuous-time stochastic process which evolves among the following three degradation levels (Figure 9.2):

- "Good": A component in this state is new or almost new (no crack is detectable by maintenance operators).

- "Medium": If the component is in this degradation state, then it is best to replace it.

- "Bad": A component in this degradation state is very likely to experience a failure in a few working hours.

A further state, "Failed," can be reached from every degradation state upon the occurrence of a shock event.

The choice of describing the degradation process by means of a small number of levels, or degradation "macro-states," is driven by industrial practice: experts usually adopt a discrete and qualitative classification of the degradation states based on qualitative interpretations of symptoms.

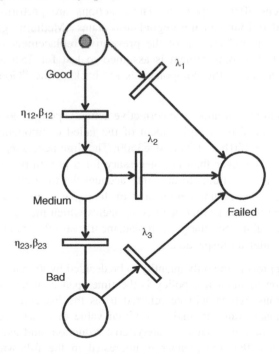

Figure 9.2   Degradation model based on a Petri net description.

This model of the degradation process can be represented by introducing the following five aleatory quantities:

- X_1 the transition time from degradation level "Good" to degradation level "Medium"

- X_2 the transition time from degradation level "Medium" to degradation level "Bad"

- X_3 the transition time from degradation level "Good" to the failed state

- X_4 the transition time from degradation level "Medium" to the failed state

- X_5 the transition time from degradation level "Bad" to the failed state.

A condition-based maintenance (CBM) approach is applied to the component. In practice, the component is periodically inspected by a plant operator according to a predefined schedule, and if during the inspection it is found in the degradation level "Medium" or "Bad," it is replaced. Replacement is also performed after component failure. The following duration and costs, in arbitrary units due to the confidentiality of the data, are assumed for the different maintenance tasks:

- Inspections: These actions aim at detecting the degradation state of the component. Inspections, which are performed periodically on the basis of a predefined schedule, are assumed to last 5 h and have an associated cost of €50.

- Replacement after inspection: These actions are performed only if the component is found in the degradation states "Medium" or "Bad" during the inspection and consist of the preventive replacement of the degraded component. The replacement is assumed to last for 25 h and costs €500. After replacement, the component is set back to the "Good" degradation level.

- Replacement after failure: The corrective action is performed after component failure and consists of replacement of the failed component. Its duration is assumed to be 100 h and its cost €3500. The time necessary for replacement after failure is longer than that necessary for a preventive replacement performed after inspections, to take into account the time elapsing between the occurrence of the failure and the start of the replacement action and the extra time needed to repair other plant components which may be damaged by the component failure. Similarly, the replacement costs after failure also are higher than those after an inspection.

Under this approach, the only quantity to be decided by the maintenance planner to fully define the maintenance policy is the time span between two successively planned inspections, which will be referred to as the inspection interval II. The maintenance planner wants to find an optimal value of II with respect to his or her performance objectives such as safety, cost reduction, and availability. In this section, we assume that the planner is interested in the following maintenance performance indicators: (i) the maintenance costs and (ii) the component downtime over its mission time. In this context, two different types of uncertainty have to be considered for the assessment of the maintenance performance indicators:

- the stochasticity of the degradation and failure process;

- the epistemic uncertainty on the parameters of the probability distributions representing the transition and failure times.

In practice, we consider a model whose input quantities are the transition and failure times, and whose output quantities are the maintenance costs and the portion of downtime over the mission time. Note that the maintenance planner is interested in the expected values of these two uncertain quantities, which will be referred to as expected maintenance costs and average unavailability over the mission time. Since the outcomes of the input quantities are subject to aleatory uncertainty described by frequentist probabilities with parameters subject to epistemic uncertainty, uncertainty propagation requires a level 2 setting (Section 6.2).

Application 9.1 (in a nutshell)

Input uncertain quantities:

- X_1 the transition time from degradation level "Good" to degradation level "Medium"

- X_2 the transition time from degradation level "Medium" to degradation level "Bad"

- X_3 the transition time from degradation level "Good" to the failed state

- X_4 the transition time from degradation level "Medium" to the failed state

- X_5 the transition time from degradation level "Bad" to the failed state

- parameters Θ_{11}, Θ_{21} of the probability distribution representing the uncertainty on X_1

- parameters Θ_{12}, Θ_{22} of the probability distribution representing the uncertainty on X_2

- parameter Θ_{13} of the probability distribution representing the uncertainty on X_3

- parameter Θ_{14} of the probability distribution representing the uncertainty on X_4.

Output quantity:

- portion U of the mission time in which the component is unavailable

- maintenance costs, C.

(continued)

> The decision maker is usually interested in the expected values of these two uncertain quantities, which are usually referred to as average unavailability, EU, over the mission time and expected maintenance costs, EC.
>
> *Type of uncertainty on the input quantities:*
>
> - aleatory on X_1, X_2, X_3, X_4, X_5
> - epistemic on $\Theta_{11}, \Theta_{21}, \Theta_{12}, \Theta_{22}, \Theta_{13}, \Theta_{14}$.
>
> *Uncertainty propagation setting:*
>
> - level 2.

9.3 Uncertainty representation

The representation of the uncertainty of the model input quantities, that is, the transition and failure times, $X_1, \ldots, X_5$, requires one to choose the probability distribution types and set their parameters.

In the context of degradation modeling, the Weibull distribution is commonly applied in fracture mechanics, especially under the weakest-link assumption (Remy *et al.*, 2010) to represent transition times between degradation states. Thus, the transition times are represented by Weibull distributions characterized by uncertain scale and shape parameters, η_{ij} and β_{ij}, respectively, for the transitions from state i toward state j ($i = \{1, 2\}$ and $j = i + 1$).

With respect to failure times X_3, X_4, X_5, their uncertainty has been represented using exponential distributions with constant failure rate λ_j, for every $j = 1, 2, 3$. The choice of using constant failure rates is driven by industrial practice: experts are familiar with this setting and comfortable with providing information about the failure rate values.

All the parameters of the distributions that model the transitions of the component among the four states of Figure 9.2 are not well known, and their evaluation comes (with imprecision) from teams of experts. With respect to the estimate of the parameters of the Weibull distributions, note that the scale parameter represents the time up to which almost 65% of the components have experienced the transition, and the shape parameters, the slopes of the Weibull probability plots. To sum up, the uncertainty situation is as follows. There are five stochastic uncertain quantities, which define the five transition times as given in Table 9.1. The distributions associated with the variables are known, and depend on a set of seven epistemic uncertain parameters, which are the shape and scale parameters of the two Weibull distributions and the failure rates pertaining to the three degradation levels.

However, the uncertainty on the third failure rate is not considered. In fact, a sensitivity analysis carried out by Baraldi, Compare, and Zio (2013a) has shown that the output of the model does not appreciably change when the value of the third failure

Table 9.1   Model parameters.

Random variables	Uncertain parameters	Description
X_1	$\Theta_1 = (\Theta_{11}, \Theta_{21})$	Transition time from degradation level "Good" to "Medium"
X_2	$\Theta_2 = (\Theta_{12}, \Theta_{22})$	Transition time from degradation level "Medium" to "Bad"
X_3	$\Theta_3 = (\Theta_{13})$	Transition time from degradation level "Good" to "Failed"
X_4	$\Theta_4 = (\Theta_{14})$	Transition time from degradation level "Medium" to "Failed"
X_5	$\Theta_5 = (\Theta_{15})$	Transition time from degradation level "Bad" to "Failed"

rate varies over a wide interval, whereas accounting for a further uncertain parameter strongly increases the computational effort.

We assume that for each uncertain parameter, three experts are involved in the assessment of its value. Each expert is asked to provide the extreme values of the interval that he or she believes contain the true value of the uncertain parameter. The intervals provided by the three experts are given in Table 9.2. The generic interval will be referred to as $J_{i,p}^{(l)}$, with $l = 1, \ldots, 3$ indicating the expert, $i = 1, \ldots, 5$, the stochastic transition or failure time, and $p = 1, \ldots, M^i$, the parameter of the probability distribution describing the transition or failure time ($M^1 = M^2 = M^3 = 2$ and $M^4 = M^5 = 1$).

According to Section 6.2.2, the information elicited from the experts has been used to build, for each uncertain parameter $\Theta_{i,p}$, $i = 1, \ldots, 5$ and $p = 1, \ldots, M^i$, an evidence space $(S_{i,p}, J_{i,p}, m_{i,p})$. The domain of the parameter $\Theta_{i,p}$, that is, the set of its

Table 9.2   Uncertainty ranges for the parameters provided by three independent sources.

Parameters		Expert knowledge					
		Expert 1		Expert 2		Expert 3	
		Min	Max	Min	Max	Min	Max
Θ_{11}	η_{12}	1843	1880	1815	1908	1720	2001
Θ_{21}	β_{12}	7.92	8.08	7.8	8.2	7.4	8.6
Θ_{12}	η_{23}	735	750	725	762	687	800
Θ_{22}	β_{23}	7.92	8.08	7.8	8.2	7.4	8.6
Θ_{13}	λ_1	$9.9 \cdot 10^{-7}$	$1.01 \cdot 10^{-6}$	$9.75 \cdot 10^{-7}$	$1.03 \cdot 10^{-6}$	$9.25 \cdot 10^{-7}$	$1.075 \cdot 10^{-6}$
Θ_{14}	λ_2	$0.99 \cdot 10^{-4}$	$1.01 \cdot 10^{-4}$	$9.75 \cdot 10^{-5}$	$1.03 \cdot 10^{-4}$	$9.25 \cdot 10^{-5}$	$1.075 \cdot 10^{-4}$
Θ_{15}	λ_3	$1 \cdot 10^{-2}$	$1 \cdot 10^{-2}$	$1 \cdot 10^{-2}$	$1 \cdot 10^{-2}$	$1 \cdot 10^{-2}$	$1 \cdot 10^{-2}$

possible values, is the union of the three intervals provided by the experts, $J_{i,p}^{(l)}$, whereas the set of focal elements $J_{i,p}$ is made up of these three intervals. Finally, the BBA assigns to each interval the same mass value

$$m_{i,p}\left(J_{i,p}^{(l)}\right) = \frac{1}{3}.$$

9.4 Uncertainty propagation

The main quantities in which the maintenance decision makers are interested are the portion of downtime in the whole mission time and the cost associated with the maintenance policy. Since both quantities are uncertain, the decision maker typically considers their expected values, that is, the mean value of the portion of downtime in the whole mission time, which is indicated by the average unavailability over the mission time, EU, and the expected costs, EC.

In order to aid the reader's comprehension of this case study, in the next we present the uncertainty propagation results obtained in the unrealistic case of no epistemic uncertainty on the probability distribution parameters, whereas in Section 9.4.2 we describe the application of the hybrid probabilistic–theory of evidence uncertainty propagation method for this case study, in the situation where all sources of uncertainty are considered.

9.4.1 Maintenance performance assessment in the case of no epistemic uncertainty on the parameters

Table 9.3 lists the values used for the parameters of the failure and degradation time probability distributions, which have been taken from Zille *et al.* (2009) and correspond to the middle point of the intervals provided by the experts (Table 9.2). In this case, propagation of the uncertainty is performed within a level 1 uncertainty propagation setting and requires the application of the Monte Carlo (MC) method (Baraldi, Compare, and Zio, 2013b).

Table 9.3   Parameters of the probability distributions.

Parameters	Nominal values
η_{12}	1861 h
β_{12}	8
η_{23}	743 h
β_{23}	8
λ_1	$10^{-6}\,\mathrm{h}^{-1}$
λ_2	$10^{-4}\,\mathrm{h}^{-1}$
λ_3	$10^{-2}\,\mathrm{h}^{-1}$

Each trial of a Monte Carlo simulation consists of generating a random walk that guides the component from one state to another, at different times. During a trial, starting from the state "Good" at time 0, we need to determine when the next transition occurs and what new state is reached by the system after the transition. The procedure, then, repeats until the time reaches the mission time. The time is suitably discretized in intervals, "bins," and counters are introduced to accumulate the contributions to unavailability. In each counter, we accumulate the time in which the component has been unavailable, that is, in the state "failed," during the bin. After all the MC histories have been performed, the content of each counter divided by the bin lengths and by the number of histories gives an estimate of the average unavailability of the component in that bin, and the average of all the obtained unavailability in the different bins gives the average unavailability of the component over the mission time. Note that this procedure corresponds to performing an ensemble average of the realizations of the stochastic process governing the system lifetime.

In this application, the mission time, $T_{\text{miss}} = 10\,000\,\text{h}$, has been divided into $N_{\text{bin}} = 20$ bins of length $500\,\text{h}$ and 5×10^4 MC simulations have been carried out. The results refer to an inspection interval of $2000\,\text{h}$. Figure 9.3 shows the values of the average unavailability of the component over the bins. The ordinate of Figure 9.3 reports the average unavailability corresponding to the different bins. In practice,

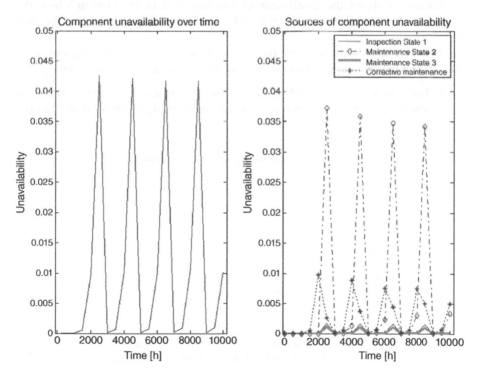

Figure 9.3 (a) Estimate of the component average unavailability over the bins partitioning the mission time; (b) identification of the different sources of unavailability.

when the MC method is applied, statistics for the time in which the component is unavailable during the bin are collected. The statistics describe how the portion of downtime is influenced by the aleatory variability associated with the stochastic model of the component behavior. The collected values of d in every bin are then averaged to get an estimate of the average unavailability over the bin (the values reported in Figure 9.3).

Since the MC method provides only an estimate of the true distribution of the unavailability in the bins, and these quantities are subject to an error, the 68.3% confidence interval is typically added to the representation of the unavailability. Note that the MC estimation error, represented by these confidence intervals, can be reduced by increasing the number of MC simulations: according to the central limit theorem (Papoulis and Pillai, 2002), the estimation error is described by a normal distribution, which tends to 0 as the number of MC simulation increases. When the standard deviation of this normal distribution is added to and subtracted from the estimated mean value, the 68.3% confidence value is then determined (Zio, 2012). Since in this case the 68.3% confidence intervals are so narrow that they seem to reduce to points. they have not been represented in the unavailability representation in Figure 9.3. This confirms that choosing to perform 5×10^4 MC simulations guarantees an acceptable precision of the results.

Figure 9.4 shows the distribution of the portion d of the time in which the component is unavailable in the bin $[2000\,\text{h}, 2500\,\text{h}]$, corresponding to the first inspection time.

The CDF in Figure 9.4 has two main steps, which can be interpreted by considering the stochastic evolution of a population of identical components:

- The first step at $d = 0.01$ is caused by the inspection of the components which have not failed before and are found in degradation state "Good."

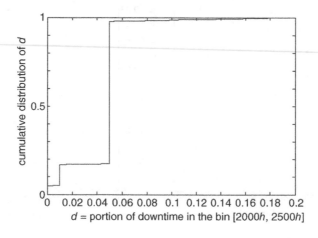

Figure 9.4   Cumulative distribution of the portion of downtime in the bin [2000 h, 2500 h] (based on Baraldi, Compare, and Zio, 2013a).

These components are unavailable during the inspection which lasts for 5 h
(= 1% of D_t).

- The second step at $d = 0.05$ is caused by CBM actions on the components
which have not failed before and are found in degradation state "Medium" or
"Bad." These components require maintenance actions that last 25 h (= 5% of
D_t).

Obviously, there are other contributions to d, which are related to:

- Replacement actions of the components failing in the previous bin, and reset
into operation in the current bin; these components cause the smoothly
increasing behavior of the CDF between $d = 0.05$ and $d = 0.2$.

- Unavailability due to maintenance actions on components that have failed in the
first bin ($[0 \text{ h}, 500 \text{ h}]$) and are thus inspected between 2000 and 2500 h.

- Unavailability due to failure of the components that have already experienced
one or more failures in the previous bins.

Notice that the downtime in the bin is always smaller than 20% of its length; this is
due to the fact that none of the components of the considered population has
experienced more than one failure in the same bin, and the duration of a replacement
action corresponds to 20% of the bin duration.

Figure 9.3(b) decomposes the average unavailability over a bin into its different
constituents: unavailability due to the inspection of the component while it is in
degradation state "Good"; unavailability due to preventive replacements if it is found
in states "Medium" or "Bad" at inspection; and unavailability due to corrective
maintenance actions that are performed upon failure.

By considering a population of components of the same type, a comparison of
Figures 9.3(a) and (b) shows that the first increase of unavailability, at $t = 2000$ h, is
mainly due to the corrective maintenance actions that replace the components that
failed within the time interval $[1500 \text{ h}, 2000 \text{ h}]$. For this purpose, notice that the scale
parameter of the Weibull distribution representing the transition from the degradation
state "Good" to the degradation state "Medium" is equal to 1861 h, and since this
quantity corresponds to the 63.21th percentile of the distribution of the transition time,
it is expected that several components will have experienced the transition toward the
state "Medium" and a small number of components will have even experienced a
further transition toward the state "Bad." The values of the failure rates associated
with these latter states (10^{-4} h^{-1} and 10^{-2} h^{-1}, respectively), which are larger than
that associated with the "Good" state (10^{-6} h^{-1}), explain the increase in the number of
components that fail in the interval $[1500 \text{ h}, 2000 \text{ h}]$.

The average unavailability over the bins shown in Figure 9.3 reaches a maximum
at $t = 2500$ h, which refers to the bin $[2000 \text{ h}, 2500 \text{ h}]$; the sources of unavailability in
this bin have been discussed above.

In the successive bins, there is an increase in the number of components whose
inspection and failure times are shifted with respect to the "crowd" (i.e., the large
number of components experiencing the same behavior), and this leads to more

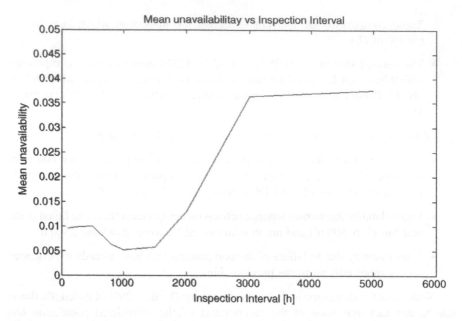

Figure 9.5 Mean unavailability (EU) corresponding to different inspection intervals.

smoothed peaks, due to replacement of components in the "Medium" degradation state, and larger average unavailability in the bins that follow these peaks, due to the replacement of components both failed and in the "Medium" degradation state. The unavailability due to replacement of components in the "Bad" state and inspections in the "Good" state remains small.

Figure 9.5 shows the average unavailability of the component in the mission time for different values of the inspection interval II. Initially, there is a decreasing behavior that reaches a minimum corresponding to $II = 1000\,h/1500\,h$; for longer inspection times, the unavailability starts to increase rapidly. This is the result of two conflicting trends: on the one hand, more frequent inspections increase the probability of finding the component in degradation states "Medium" and "Bad"; this prevents the component from failing and thus avoids the corresponding long replacement time after failure. On the other hand, frequent replacements are ineffective, since component life is not completely exploited in this case. The minimum at $II = 1500\,h$ represents the optimal balance between these two tendencies.

9.4.2 Application of the hybrid probabilistic–theory of evidence uncertainty propagation method

We now consider the epistemic uncertainty on the parameters of the failure and degradation time distributions. The uncertainty propagation method described in Section 6.2.2 is applied to this level 2 uncertainty setting. The output of the model has been defined as a vector $\mathbf{Z}$ formed by $N_{bin} + 2$ elements: the first N_{bin} elements represent the portions of downtime in the corresponding bins, element $N_{bin} + 1$ represents the portion of downtime in the whole mission time, and the last bin represents the cost associated with the maintenance policy.

As describe in Section 6.2.2, the uncertainty propagation method provides summary measures of the model output quantities. In this work, we focus on the mean values of the output quantities, that is, the average unavailability EU over the time bins that the mission time has been divided into, and the expected costs EC. Considering, for example, the average unavailability EU over the mission time, the method provides the belief and plausibility measures $Bel(A)$ and $Pl(A)$ for any interval of possible average unavailability values, $A \subset [0, 1]$. The results will be illustrated considering intervals $[0, u)$ with u in $[0, 1]$, and the measures $Bel([0, u))$ and $Pl([0, u])$ will be referred to as belief and plausibility distributions.

The application of the uncertainty propagation method requires a fixed number of samples, M_e, to be generated from the space of the uncertain parameters subject to epistemic uncertainty. Note that the larger the value of M_e, the larger the number of output values, and, thus, the higher the precision in the identified pair of distributions $[Bel, Pl]$. Therefore, setting M_e requires a trade-off between the precision of the distributions and the need to reduce the computational time. In this work, M_e has been set to 2000.

9.5 Results

Figure 9.6 shows the obtained plausibility and belief distributions of the average unavailability over the different bins into which the mission time has been divided. In the first bins (i.e., from $t = 500$ h to $t = 1500$ h), the plausibility and belief distributions are very close to each other and reach 1 in correspondence to a value of the average unavailability very close (or even equal) to 0; this is due to the fact that the average bin unavailability tends to remain very small for any combination of the values of the epistemic uncertain parameters ranging in the intervals provided by the experts.

The situation is different at $t = 2000$ h, when both plausibility and belief distributions are shifted toward higher values of the unavailability. This is due to an increase in the number of components that experience a failure in the bin $[1500 \text{ h}, 2000 \text{ h}]$, due to components' transitions toward the degradation states "Medium" and "Bad," as explained above in the case without epistemic uncertainty (Section 9.4.1). Note also that the gap between the plausibility and belief distributions is quite large, when compared to those of the first bins. This is due to the fact that the behavior of the components is greatly influenced by the particular combination of the uncertain parameters. For example, considering that the scale parameter represents the 63rd percentile, a combination of the values $\eta_{12} \cong 720$ h and $\eta_{23} \cong 690$ h leads to simulated histories in which it is very likely that the components experience a failure before $t = 2000$ h, with the consequent large unavailability; on the contrary, the combination $\eta_{12} \cong 2000$ h and $\eta_{23} \cong 800$ h results in histories in which a failure in the bin $[1500 \text{ h}, 2000 \text{ h}]$ rarely occurs.

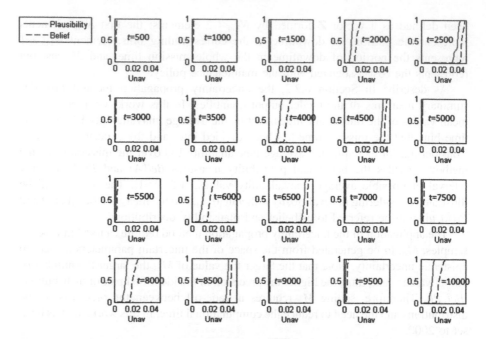

Figure 9.6   Plausibility and belief distributions of the mean values of the unavailability over time, obtained by the hybrid method (based on Baraldi, Compare, and Zio, 2013a).

In the next bin, at $t = 2500$ h, the distributions are even more shifted toward the right part of the unavailability axis; this is in agreement with the behavior of the unavailability in the case with no uncertainty on the model parameters (Figure 9.3). In the successive bins (Figure 9.6), the plausibility and belief distributions follow the "cycle" of the first bins; for example, the curves relevant to the bin [1500 h, 2000 h] are similar to the corresponding ones in the bin [3500 h, 4000 h]; the differences between the plausibility distributions and the belief distributions pertaining to "similar" bins are due to an increase in the number of components that experience a life different from that of the "crowd," as explained previously in Section 9.4.1.

In an attempt to summarize the distributions in Figure 9.6, Figure 9.7 shows lower and upper bounds of the median of the average unavailability over the bins. The median represents the 50th percentile of the average unavailability in the bins. For comparison, the estimates of the average unavailability over the bins found in Section 9.4.2, where the epistemic uncertainty was not considered, are also provided in Figure 9.7.

Figures 9.8 and 9.9 show the plausibility and belief distributions of the mean unavailability and cost over the mission time, respectively, corresponding to three different values of the *II*, namely *II* = 1000 h, *II* = 1500 h, and *II* = 2000 h. The small degree of uncertainty in the values of both unavailability and costs, when the component is inspected every 1000 h, derives from the fact that the "crowd" remains

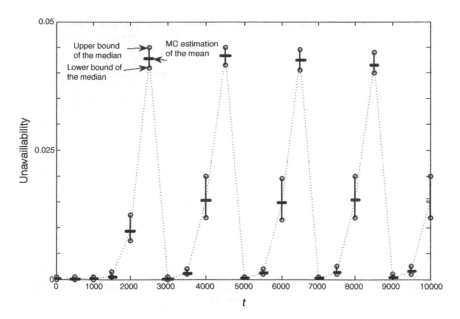

Figure 9.7   Lower and upper bounds of the median of the average unavailability over the bins (based on Baraldi, Compare, and Zio, 2013a).

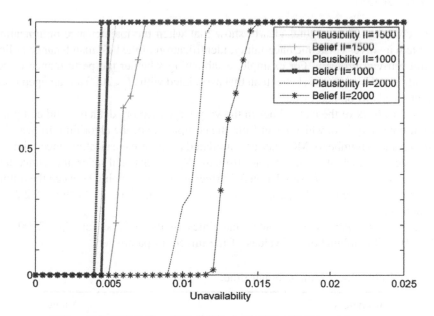

Figure 9.8   Plausibility and belief distributions of the mean unavailability over the time horizon, for different values of the control variable II (based on Baraldi, Compare, and Zio, 2013a).

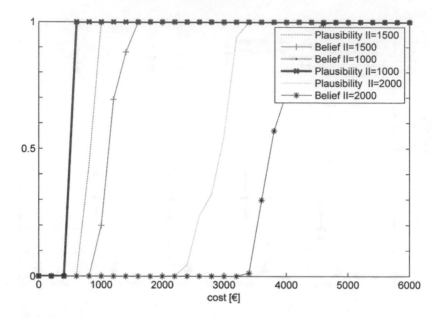

Figure 9.9 Plausibility and belief distributions of the expected cost EC over the time horizon, for different values of the control variable Π (based on Baraldi, Compare, and Zio, 2013a).

very compact. These figures clearly show that when the maintenance optimization problem involves epistemic uncertainty, identification of the best maintenance policy is not a trivial problem. For example, establishing whether the performance corresponding to $II = 1500$ h is better than that associated with $II = 2000$ h is an open issue which needs to be addressed.

A drawback of the method lies in the very large memory demands and computational times required, which result from the complexity of the algorithm. In fact, this requires that a number of MC trials are simulated to capture the aleatory uncertainty of the system for each of the M_e samples from the N_u-dimensional space of the uncertain parameters (i.e., $N_S \times N_T$ simulations). Furthermore, the mapping between the output and the N_u-dimensional space (step 5 of the procedure in Section 5.2.2) is burdensome.

Table 9.4 reports the computational times of the method for $N_T = 2000$ and $N_S = 10\,000$ combinations of values of the uncertain parameters.

Table 9.4   Computational time.

Parameters	Values
Number of MC trials	2000
Number of combinations of uncertain parameters	8000
CPU time (Intel Core 2 duo, 3.17 GHz, 2GB RAM)	$\approx$30 h

10

Uncertainty representation and propagation in event tree analysis

In this chapter, we illustrate the application of the uncertainty representation and propagation methods to event tree analysis in the nuclear industry. The application is in part based on Baraldi and Zio (2008).

10.1 Event tree analysis

Event tree analysis (ETA) is an inductive logic method for identifying the various accident sequences which can result from a single initiating event and is based on the discretization of the real accident evolution into a few macroscopic events. The accident sequences are then quantified in terms of their probability of occurrence. In the following, we give only the very basic principles of the technique. The interested reader is invited to look at the specialized literature for further details, for example, Zio (2007) and Henley and Kumamoto (1992), and references therein, from which the material here has been synthesized.

ETA begins with the identification of the initiating event, which is typically a component failure or an external failure. Then, all the safety functions which are intended to mitigate the accident are defined and organized according to their logic of intervention. For example, Figure 10.1 shows a graphical example of an event tree: the initiating event is depicted by the initial horizontal line and the system states are then connected in a stepwise, branching fashion: system success and failure states have been denoted by S and F, respectively. The accident sequences that result from the tree structure are shown in the last column. Each branch yields one particular accident sequence; for example, IS_1F_2 denotes the accident sequence in which the initiating event (I) occurs, system 1 is

Uncertainty in Risk Assessment: The Representation and Treatment of Uncertainties by Probabilistic and Non-Probabilistic Methods, First Edition. Terje Aven, Piero Baraldi, Roger Flage and Enrico Zio.
© 2014 John Wiley & Sons, Ltd. Published 2014 by John Wiley & Sons, Ltd.

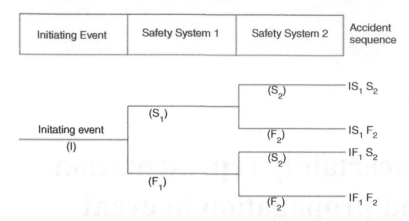

Figure 10.1   Illustration of system event tree branching (Reliability Manual for Liquid Metal Fast Reactor (LMFBR) Safety Programs, 1974) (based on Zio, 2007).

called upon and succeeds (S_1), and system 2 is called upon but fails to perform its defined function (F_2). Note that the system states on a given branch of the event tree are conditional on the previous system states having occurred. With reference to the previous example, the success and failure of system 1 must be defined under the condition that the initiating event has occurred; likewise, in the upper branch of the tree corresponding to system 1 success, the success and failure of system 2 must be defined under the conditions that the initiating event has occurred and system 1 has succeeded.

Once the final event tree has been constructed, the last task is to compute the probabilities of the various accident sequences. This phase requires knowledge of the probabilities of occurrence of the single events in the tree, that is, the conditional probability of the occurrence of the event, given that the events which proceed on that sequence have occurred. Multiplication of the conditional probabilities for each branch in a sequence gives the probability of that sequence.

10.2   Case study

Accident conditions in a nuclear power plant are typically classified as design basis accidents, that is, accident conditions against which the plant is designed according to established design criteria, and for which damage to the fuel and the release of radioactive materials are kept within authorized limits. Accidents less severe or as severe as design basis accidents are called "within design basis accidents" and those more severe are called "beyond design basis accidents" (Figure 10.2). In this latter category, those accidents involving significant core degradation which can potentially have catastrophic consequences in case of the release of radioactive products into the environment are called "severe accidents" and are the subject of a lot of attention from plant designers, owners, regulatory bodies, and research communities.

In this context, we consider one of the most feared events in the nuclear power plant industry: an anticipated transient without scram (ATWS) (Huang, Chen, and

Accident conditions

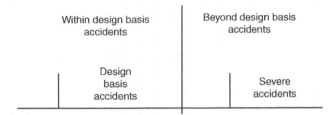

Figure 10.2   Nuclear accident classification according to IAEA safety glossary (based on IAEA Safety Glossary, 2007).

Wang, 2001). We aim at quantifying the probability of having a severe accident as a consequence of this event.

An ATWS occurs if the protection system diagnoses potential damage to the plant requiring the insertion of the control rods into the reactor core to shut down the nuclear reaction, but the operation is unsuccessful. To analyze the sequence of occurrences following an ATWS, the event tree in Figure 10.3 is considered, from Taiwan's nuclear power plant II operating PRA draft report (Nuclear Energy Research Center, 1995). The headings (branching events) of the tree are described in Table 10.1.

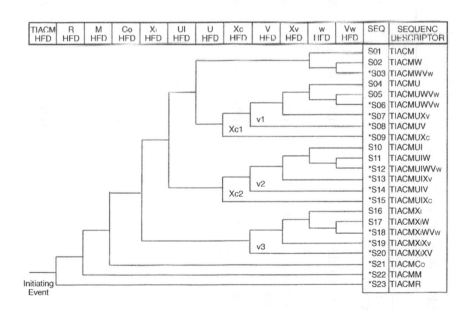

*Figure 10.3   The event tree (Huang, Chen, and Wang, 2001). The upper branch corresponds to the non-occurrence of the event, the lower branch to the occurrence; SEQ = sequence number, * = severe consequence accident (based on Huang, Chen, and Wang, 2001).*

Table 10.1 Event tree headings (top events) (Huang, Chen, and Wang, 2001); ν = probability of occurrence, HFD = Hardware-Failure-Dominated, HED = Human-Error-Dominated.

Event	Acronym	Type	ν	Description
Main condenser isolation ATWS	T1ACM	HFD	ν_1	This event happens when the reactor is isolated and the automatic scram system fails. It is also assumed that mechanical failures cannot be repaired within the allowable time
Recirculation pump trip	R	HFD	ν_2	If the plant fails to scram, an automatic recirculation pump system is required to limit power generation immediately. A failure of the automatic recirculation pump system results in event R
Safety/relief valves (S/RVs) open	M	HFD	ν_3	At the time the reactor is isolated, at least 13 of 16 S/RVs must open to prevent overpressurization of the reactor vessel. If insufficient S/RVs open, then event M happens
Boron injection	C_0	HFD	ν_4	When an ATWS event happens, the power of the core is very high. If the power cannot be slowed down to the state of shutdown, and the vapor produced by the reactor continues to inject into the suppression pool, then the temperature increases to fail the high-pressure system. This increases the possibility of core meltdown. As a result, the automaticredundant reactivity control system (RRCS) is supposed to inject liquid boron into the vessel to shut down the reactor safely. If automatic RRCS fails, and operators fail to inject liquid boron by using thestandby liquid control system (SLCS), it results in event C_0. It is assumed that operators cannot manually inject liquid boron within the allowable time
ADS inhibit	X_1	HED	ν_{12}	The automatic depressurization system (ADS) is designed to decrease the pressure of the reactor in order to start the

low-pressure system. The low-pressure system injects water into the reactor vessel to protect the fuel. When an ATWS event happens, the reactor power is controlled by the level of water in the core. Since high-level water causes high power, the operator should inhibit all ADS valves manually. If the operator fails to do so, event X_I occurs

Early high-pressure makeup	$U1$	HFD	ν_5	Following the stop of feedwater supply, the high-pressure makeup system is supposed to work automatically when an automatic actuation alarm appears as soon as the water level is lowered till level 2. The water level is expected to reach the top of the fuel. Thus, if the high-pressure system fails to work automatically, it leads to event $U1$
Long-term high-pressure makeup	U	HFD	ν_6	The success criterion of avoidance of this event is that the high-pressure system can maintain the water level in the vessel 24 h after the start. If the system fails and causes event U, then using the low-pressure system to maintain the water level is needed
Manual reactor depressurization	$X_C(X_{C1}, X_{C2})$	HED	ν_{13}, ν_{14}	If the pressure in the reactor vessel is too high to set up the low-pressure system, the operator should depressurize the vessel manually in time to avoid core meltdown. Due to the different conditional probabilities of occurrence of this event in different accidental sequences, X_C is called X_{C1} in sequences 4–9 characterized by the non-occurrence of event U_1 and $XC2$ in sequences 10–15 characterized by the occurrence of event U_1
Reactor inventory makeup at low pressure	$V(V_1, V_2, V_3)$	HFD	ν_7, ν_8, ν_9	If the low-pressure system fails as well as the high-pressure system, then event V occurs and the water level in the vessel will be so low as to probably cause core meltdown. Due to the different conditional probabilities of occurrence of this event in *(continued)*

Table 10.1 (Continued)

Event	Acronym	Type	ν	Description
				different accidental sequences, V is called V_1 in sequences 4–7, V_2 in sequences 10–14, and V_3 in sequences 16–20
Vessel overfill prevention	X_V	HED	ν_{15}	When the pressure in the vessel is decreased till the level is low enough for the low-pressure system to inject water, a huge amount of water comes into the core. The operator should pay attention to the water level and make sure that the level is kept not so high as to lead to core meltdown. The definition of this event is the operator fails to complete this job
Long-term heat removal	W	HFD	ν_{10}	The residual heat removal (RHR) system is initialized to cool down the suppression pool and containment in order to ensure other supporting systems work well. If this system fails, event W happens
Vessel inventory makeup after containment (CTMT) failure	V_W	HFD	ν_{11}	The CTMT might fail because of overpressure or overheat. Water supply in the reactor vessel must be kept to protect the fuel from not melting in the condition of CTMT failure. Among these events, X_I, X_C, and X_V are mainly caused by human error. The others are mainly caused by hardware failures

According to the assumptions made (PRA Report, 1995), the frequency probabilities of occurrence of events X_C and V are conditioned by the occurrence of events U_1 and X_1, whereas the frequency probabilities of events X_v, W, and V_W are considered as constants in the different sequences.

The quantities of interest are the frequency probabilities of occurrence of the 23 identified accident sequences and of a severe accident (Figure 10.3). These quantities are quantified on the basis of the frequency probability of occurrence of the single events.

Due to our lack of knowledge, the true value of the frequency probabilities of occurrence of the single events are unknown. Thus we need (i) to represent the epistemic uncertainty to which these probabilities are affected and (ii) to properly propagate these epistemic uncertainties into the probability of occurrences of the single sequences of events and of a severe accident.

With respect to the mathematical formulation of the problem, we consider the models g_r, $r = 1, \ldots, 23$, and compute the frequency probability of occurrence of the rth accident sequence (output of the model), p_{Seq_r}, from the frequency probabilities ν_i of occurrence/non-occurrence of the single events along the sequence (input of the model), according to

$$p_{Seq_r} = g_r(\nu_1, \nu_2, \ldots, \nu_{15}) = \prod_{i \text{ occurs in } Seq_r} \nu_i \prod_{i \text{ does not occur in } Seq_r} (1 - \nu_i), \quad r = 1, \ldots, 23.$$

$$(10.1)$$

For an interpretation of the quantified frequency probability, see Example 2.2 in Section 2.2.

Finally, the probability of occurrence of a severe accident is obtained by summing the frequency probabilities of all the sequences that lead to severe consequences:

$$p_{Sev} = \sum_{r:Seq_r \text{ is Sev}} p_{Seq_r}. \qquad (10.2)$$

Application 10.1   (in a nutshell)

Input uncertain quantities:

- frequency probabilities of occurrences of the branching events, ν_i, $i = 1, \ldots, 15$ (Table 10.1).

Output quantities:

- frequency probabilities of occurrences of the 23 identified accident sequences, p_{Seq_r}, $r = 1, \ldots, 23$
- frequency probability of occurrences of severe consequences, p_{Sev}.

(*continued*)

Models:

$$P_{Seq_r} = g_r(\nu_1, \nu_2, \ldots, \nu_{15}) = \prod_{i \text{ occurs in } Seq_r} \nu_i \prod_{i \text{ does not occur in } Seq_r} (1 - \nu_i), \quad r - 1, \ldots, 23$$

$$P_{Sev} = \sum_{r: Seq_r \text{ is Sev}} P_{Seq_r}.$$

Type of uncertainty on the input quantities:

- ν_i, $i = 1, \ldots, 15$, are subject to epistemic uncertainty.

10.3 Uncertainty representation

According to Huang, Chen, and Wang (2001), the events in Table 10.1 can be distinguished into 11 hardware-failure-dominated (HFD) events and 4 human-error-dominated (HED) events, and different sources of information are available for the representation of the uncertainties on their frequency probabilities of occurrence, namely:

- experimental data for the HFD events; and

- expert knowledge for the HED events.

With respect to the HFD events, the uncertainties on their frequency probabilities of occurrence, ν_i, $i = 1, \ldots, 11$, have been represented using the lognormal probability distributions $p_{\nu_i}(\nu_i)$ shown in Figure 10.4, whose means and standard deviations are given in Table 10.2. Lognormal probability distributions are largely used in risk assessments to represent the uncertainty on the frequency probability of events, and the parameters of the lognormal distributions can be easily assessed using lognormal probability plots (Burmaster and Hull, 1997).

On the contrary, with respect to the probability of the HED events, ν_i, $i = 12, 13, 14, 15$, experimental data are not available and we rely on the knowledge of four experts. In particular, expert knowledge has been elicited in the form of trapezoidal possibility distributions $\pi_{\nu_i}(\nu_i)$, $i = 12, 13, 14, 15$; see Figure 10.5. Details of the method used for eliciting the possibility distributions and the aggregation of the different expert opinions can be found in Huang, Chen, and Wang (2001).

Note the very different situations of knowledge of the expert on the probabilities of occurrence ν_{13} and ν_{14}. The probability of occurrence of event 14, ν_{14}, is known only with very large imprecision: the core is very large since all the values of ν_{14} between 0.1 and 1 have an associated possibility distribution $\pi(\nu_{14})$ equal to 1; that is, they are regarded by the experts as wholly possible. On the contrary, the probability of occurrence of event 13, ν_{13}, is very well defined: the support of ν_{13} is limited in the interval [0.0015, 0.0082].

Table 10.2   Parameters of the probability density functions (Huang, Chen, and Wang, 2001).

Event	PDF	Median	Error factor
T1ACM	$p_{\nu_1}(\nu)$	$1.52 \cdot 10^{-7}$	8.42
R	$p_{\nu_2}(\nu)$	$1.96 \cdot 10^{-3}$	5.00
M	$p_{\nu_3}(\nu)$	$1.00 \cdot 10^{-5}$	5.00
C_0	$p_{\nu_4}(\nu)$	$1.37 \cdot 10^{-2}$	3.00
U1	$p_{\nu_5}(\nu)$	$8.45 \cdot 10^{-2}$	3.00
U	$p_{\nu_6}(\nu)$	$2.13 \cdot 10^{-3}$	5.00
V_1	$p_{\nu_7}(\nu)$	$1.12 \cdot 10^{-6}$	10.00
V_2	$p_{\nu_8}(\nu)$	$3.40 \cdot 10^{-6}$	10.00
V_3	$p_{\nu_9}(\nu)$	$9.49 \cdot 10^{-5}$	10.00
W	$p_{\nu_{10}}(\nu)$	$2.03 \cdot 10^{-5}$	10.00
V_W	$p_{\nu_{11}}(\nu)$	$4.00 \cdot 10^{-1}$	2.40

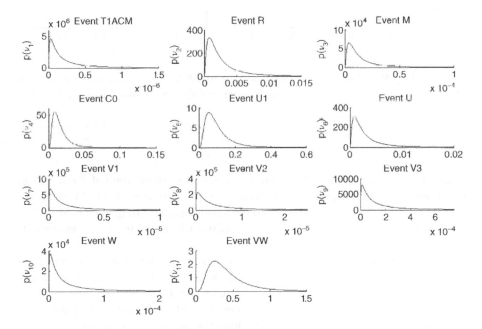

Figure 10.4   Probability density functions for the probabilities of occurrence of the 11 HFD events of Table 10.1 (based on Baraldi and Zio, 2008).

10.4   Uncertainty propagation

As some uncertain quantities are represented using probability distributions and others using possibility distributions, the hybrid probabilistic–possibilistic uncertainty propagation method described in Section 6.1.2 has been applied. In

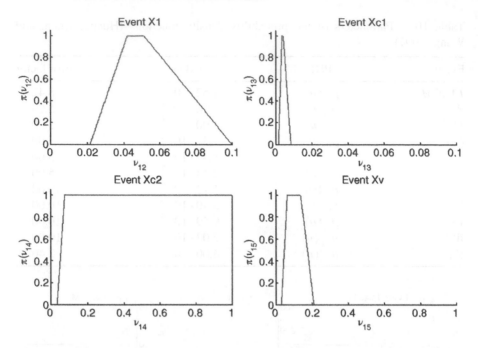

Figure 10.5   Possibility distributions of the probabilities of occurrence of the four HED events of Table 10.1 (based on Baraldi and Zio, 2008, and Huang, Chen, and Wang, 2001).

practice, the uncertainty propagation method has been applied to a model providing 24 output quantities (the frequency probabilities of the 23 identified accident sequences, p_{Seq_r}, in (10.1) and the severe accident frequency probability, p_{Sev}, in (10.2)) on the basis of 15 input quantities (the event frequency probabilities, ν_i).

The uncertainty propagation method provides, for the frequency probabilities of occurrence of each sequence and for the severe accident sequence frequency probability, the belief and plausibility measures, $Pl(A)$ and $Bel(A)$, for a generic interval of values, A. Considering for example $A = [0, u)$, the belief and plausibility measures $Pl(A)$ and $Bel(A)$ provide a description of the uncertainty that the frequency probability of the sequence is lower than u.

With respect to setting the model parameters, an acceptable trade-off between the precision of the estimate of the output distributions and computational time has been found using $M = 1000$ random realizations of the probabilistic variables and 21 α-cuts of the possibility distributions. It has been verified that increasing M or the number of α-cuts will result in linearly increasing the computational burden without any remarkable advantage for the precision of the output distributions.

10.5   Results

Figure 10.6 shows the belief and plausibility functions of the set $[0, u)$ obtained for the frequency probabilities of occurrence of sequences 13, 15, 22 and for the severe consequence accident. The three sequences have been chosen because they represent distinctly interesting cases of uncertainty propagation; on the other hand, the probability of a severe consequence accident is an important quantity for the evaluation of the risk.

In the top graph, notice that $Bel([0, u))$ (lower curve) and $Pl([0, u))$ (upper curve) of the probability of sequence 13 are quite far from one another, indicating large imprecision. This is mainly because sequence 13 is characterized by the non-occurrence of the HED event X_{C2} that is known only with very large imprecision (bottom left graph in Figure 10.5). For example, the plausibility measure that sequence 13 will never occur, that is, $Pl(p_{Seq_{13}} = 0)$ is equal to 1, while its belief measure, $Bel(p_{Seq_{13}} = 0)$, which represents its necessity, is equal to 0. The plausibility measure is 1 due to the fact that the possibility of having $\nu_{14} = 1$ is 1. That is, the expert believes that it is possible that event X_{C2} always occurs; on the contrary, event X_{C2} appears in sequence 13 as not occurring, so that it is plausible that sequence 13 never occurs, that is, $Pl(p_{Seq_{13}} = 0) = 1$.

The opposite occurs for sequence 22, which is characterized by the occurrence of only HFD events. In this case, the uncertainty of the input quantities, ν_1 and ν_2 of function g_{22}, are represented by probability distributions and, thus, $Bel([0, u))$ and $Pl([0, u))$ coincide and can be interpreted as the cumulative distribution $F(u)$ (third graph from the top in Figure 10.6).

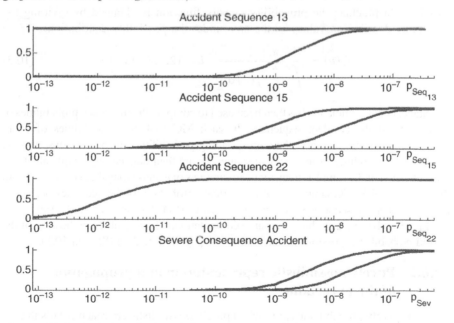

Figure 10.6   Belief and plausibility measures of the set $[0, u)$ resulting from application of the hybrid approach (based on Baraldi and Zio, 2008).

An intermediate situation between the cases of sequences 13 and 22 is represented by sequence 15 (second graph from the top of Figure 10.6).

The results of uncertainty propagation by the hybrid approach have been compared to those obtained by a purely probabilistic and a purely possibilistic uncertainty representation and propagation approach.

10.6   Comparison of the results to those obtained by using other uncertainty representation and propagation methods

10.6.1   Purely probabilistic representation and propagation of the uncertainty

For the application of a probabilistic approach to the propagation of the uncertainty in the frequency probabilities of occurrence of the branching events to the frequency probabilities of occurrence of the accident sequences, one requires the availability of probability density functions representing the epistemic uncertainty on the frequency probability of occurrence of each single event. To this end, the possibility distributions describing the uncertainty in the frequency probabilities of occurrence of the HED events (Figure 10.5) are transformed into probability density functions. This has been achieved, in this application, according to the technique presented by Yager (1996) for discrete variables and here extended to the case of continuous variables. In practice, the probability density function is obtained by dividing the possibility distribution values by the area under the possibility distribution:

$$p_{\nu_i}(\nu) = \frac{\pi_\nu(\nu)}{\int_0^{+\infty} \pi_\nu(\nu)\, d\nu} \qquad i = 12,\ 13,\ 14,\ 15. \tag{10.3}$$

Standard MC simulation has then been used to compute the frequency probabilities of occurrence of the accident sequences. In each MC trial, the probabilities of event occurrences are sampled from the corresponding probability density functions, and the frequency probabilities of the accident sequences are then obtained by simple algebraic multiplication of the sampled frequency probabilities of events along the sequences. From the repetition of M MC sampling trials, empirical distributions of the frequency probabilities of the accident sequences are obtained (Figure 10.7). The number of MC samplings used is $M = 10^5$, so as to have the same computational time obtained in the case of the hybrid probabilistic–possibilistic uncertainty propagation method (Section 10.5).

10.6.2   Purely possibilistic representation and propagation of the uncertainty

As in the purely probabilistic approach, a purely possibilistic approach entails that all input quantity uncertainties be described in terms of possibility distributions, including the HFD event parameters.

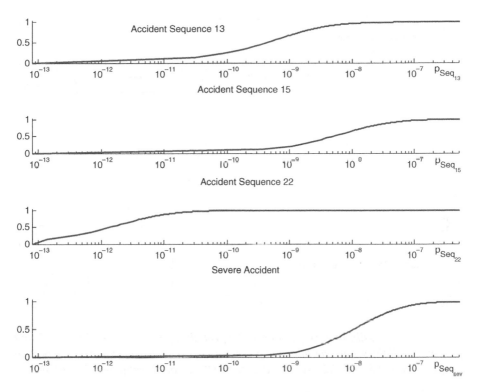

Figure 10.7   Cumulative distributions of the probabilities of accident sequences 13, 15, 22 and of a severe consequence accident resulting from the probabilistic approach (based on Baraldi and Zio, 2008).

According to Huang, Chen, and Wang (2001), the probability density functions that describe the uncertainty about the frequency probabilities of the HFD events have been transformed into triangular possibility distributions, peaked (i.e., with possibility distribution equal to 1) at the mean value of the corresponding probability density functions, and with the two vertexes of the base (possibility distribution equal to 0) positioned corresponding to the 5% and 95% percentiles. The probabilities of occurrence of the accident sequences are then computed by using the extension principle of fuzzy set theory (6.7). In particular, only the 20 α-cuts (0.05,0.1, . . . ,1) reported in Huang, Chen, and Wang (2001) have been considered, together with the zero α-cuts (the supports of the possibility distributions) which are not given in Huang, Chen, and Wang (2001) but have been obtained here by simple interpolation of the available α-cut values.

The obtained possibility distributions of the frequency probabilities of occurrence of sequences 13, 15, 22 and of a severe consequence accident are displayed in Figure 10.8.

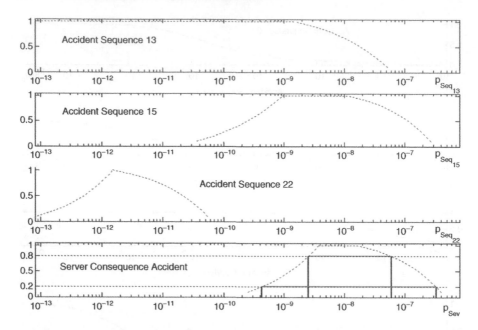

Figure 10.8 Possibility distributions of the probability of sequences 13, 15, 22 and (bottom) of all the sequences that lead to a severe consequence accident. In the bottom graph, α-cuts 0.8 and 0.2 are also depicted. (Based on Baraldi and Zio, 2008.)

To compare these results to the previous ones, the necessity and possibility measures, $N([0, u))$ and $\Pi([0, u))$, which correspond to the belief and plausibility measures $Bel([0, u))$ and $Pl([0, u))$ of the hybrid probabilistic–possibilistic approach, can be obtained by using (4.1) and (4.2). Figure 10.9 shows the distributions obtained. Notice that Figures 10.8 and 10.9 present the same information from two different points of view: the possibility distributions of the former figure can be read in terms of α-cuts, while the belief and plausibility measures of the latter figure can be interpreted by fixing a level of credibility and providing an interval of possible values for the one-sided upper limit by taking the upper and lower bounds from $Bel([0, u))$ and $Pl([0, u))$. For example, the α-cut 0.8 of the severe accident probabilities is $(2.549 \cdot 10^{-9}, 6.056 \cdot 10^{-8})$ and the α-cut 0.2 is $(4.261 \cdot 10^{-10}, 3.322 \cdot 10^{-7})$, whereas the 80% one-sided upper limit of the severe consequence accident probabilities lies within $(2.549 \cdot 10^{-9}, 3.322 \cdot 10^{-7})$. Thus, the $a/100$ one-sided upper limit can have values between the left limit of the α-cut a and the right limit of the α-cut $(1 - a)$. This is due to the definition of plausibility and belief measures given in Section 10.4, from which we get that the value u_1 is such that $Pl([0, u_1)) = a$ corresponds to the left limit of the α-cut a, that is, the minimum value of X with a possibility of a, while the value u_2 is such that $Bel([0, u_2)) = a$ corresponds to the right limit of the α-cut $(1 - a)$. In particular, the belief measure reaches 1 corresponding to the right limit of the support.

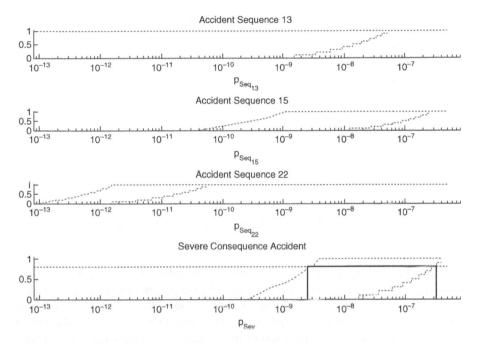

Figure 10.9 Belief $\mathrm{Bel}([0, \mathrm{u}))$ *and plausibility* $\mathrm{Pl}([0, \mathrm{u}))$ *measures obtained from the possibility distributions of Figure 10.8. The interval of possible values for the 0.8 upper credibility bound is also depicted. (Based on Baraldi and Zio, 2008.)*

10.7   Result comparison

Figures 10.10 to 10.13 show the comparison of the cumulative distributions obtained by the purely probabilistic approach to the belief and plausibility measures obtained by the possibilistic and hybrid approaches.

10.7.1   Comparison of results

10.7.1.1   Comparison between the hybrid and the purely probabilistic approach

In the case of the computation of the probability of sequence 22, characterized by input parameters whose uncertainty is represented only by probabilistic distributions, the hybrid and probabilistic approaches coincide. The small differences between the cumulative distribution obtained by the probabilistic approach and the belief and plausibility measures in Figure 10.12 are due to the different numbers of samplings M that have been performed for the probabilistic estimation ($M = 10^5$) and for the hybrid approach estimation ($M = 10^4$).

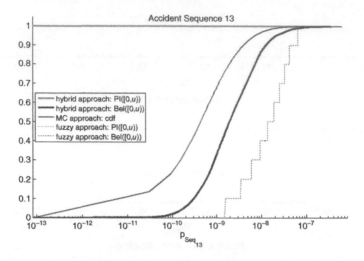

Figure 10.10   Comparison of the cumulative distribution of the probability of occurrence of accident sequence 13 obtained by the purely probabilistic approach to the belief and plausibility measures, $\mathrm{Bel}([0,\mathrm{u}))$ *and* $\mathrm{Pl}([0,\mathrm{u}))$, *obtained by the purely possibilistic and hybrid approaches (the measures obtained by the purely possibilistic and the hybrid approaches are in this case completely overlapping). (Based on Baraldi and Zio, 2008.)*

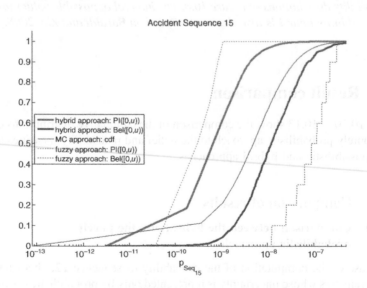

Figure 10.11   Comparison of the cumulative distribution of the probability of occurrence of accident sequence 15 obtained by the purely probabilistic approach to the belief (lower curve) and plausibility (upper curve) measures, $\mathrm{Bel}([0,\mathrm{u}))$ *and* $\mathrm{Pl}([0,\mathrm{u}))$, *obtained by the purely possibilistic and hybrid approaches (based on Baraldi and Zio, 2008).*

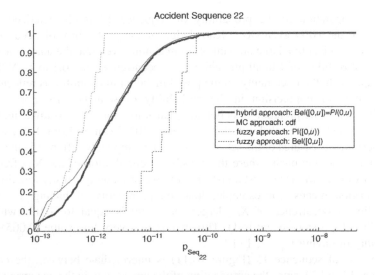

Figure 10.12   Comparison of the cumulative distribution of the probability of occurrence of accident sequence 22 obtained by the purely probabilistic approach to the belief (lower curve) and plausibility (upper curve) measures, $\mathrm{Bel}([0, \mathrm{u}))$ *and* $\mathrm{Pl}([0, \mathrm{u}))$, *obtained by the purely possibilistic and hybrid approaches (based on Baraldi and Zio, 2008).*

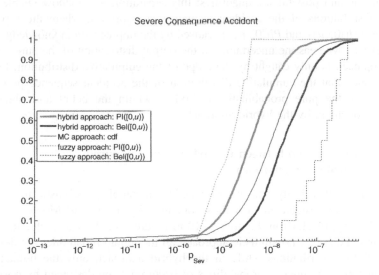

Figure 10.13   Comparison of the cumulative distribution of the probability of occurrence of a severe consequence accident obtained by the probabilistic uncertainty propagation approach to the belief (lower curve) and plausibility (upper curve) measures, $\mathrm{Bel}([0, \mathrm{u}))$ *and* $\mathrm{Pl}([0, \mathrm{u}))$, *obtained by the possibilistic and hybrid approaches (based on Baraldi and Zio, 2008).*

The computation of the probability of sequence 13 is characterized by the presence of the HED event X_{C2}. Knowledge of the probability of occurrence of this event is subject to a large amount of imprecision given that the analyst assigns a degree of possibility of 1 to all probabilities of occurrence from 0.1 to 1. Within the hybrid approach, the uncertainty on the possibilistic input quantities is reflected by a large amount of imprecision in the probability of sequence 13. For example, according to the interpretation of belief and plausibility measures as lower and upper probability limits, the probability that $p_{Seq_{13}} < 1 \cdot 10^{-13}$ can go from 0 to 1 (believe and plausibility measures of the set $[0.1 \cdot 10^{-13})$). Completely different is the case of the probabilistic approach, where the possibility distribution of the X_{C2} frequency probability of occurrence is first transformed into a probability density function. This transformation means, for example, that the probability of having a value of probability of occurrence of X_{C2} larger than 0.99 is equal to 0.011 (while the possibility of the same event is 1). This results in a very low value (0.005) of the probability of having $p_{Seq_{13}} < 1 \cdot 10^{-13}$.

The case of sequence 15 (Figure 10.11) is intermediate between the cases of sequences 13 and 22, since the propagation of the uncertainty in the occurrence of the HED event X_{C2} still causes a separation between the belief and the plausibility measures, but in reduced terms if compared to the case of sequence 13.

To sum up the comparison, if at least one input parameter is described by a possibility distribution (sequences 13 and 15), the hybrid approach explicitly propagates the uncertainty by partially separating the contributions coming from the probabilistic and possibilistic quantities; this separation is somehow visible in the output distributions of the accident sequence probabilities, where the separation between $Bel([0, u))$ and $Pl([0, u))$ is caused by the imprecision in knowledge of the HED events, whereas the uncertainty in the output distribution of the pure probabilistic approach is only caught by the slope of the cumulative distribution. Furthermore, notice that the cumulative distribution of the accident sequence probability obtained by the pure probabilistic method is within the belief and plausibility measures obtained by the hybrid approach.

10.7.1.2   Comparison between the hybrid and the purely possibilistic approach

The belief and plausibility measures obtained by the purely possibilistic approach are in all cases more separated than those obtained by the hybrid Monte Carlo and possibilistic approach (from Figures 10.10 to Figures 10.13). This is due to the fact that within the purely possibilistic approach all the input probabilities are described by possibility distributions while in the hybrid approach only the human-error-dominated event frequency probabilities of occurrence are described by possibility distributions. In particular, notice that the difference between the belief and plausibility curves obtained by the two approaches is more remarkable in the case of sequence 22 characterized only by hardware-failure-dominated events than in the case of sequence 13, in which the uncertainties due to the human-error-dominated events are predominant. The probability of occurrence of Sequence 15 (Figure 10.11) is

Table 10.3   Uncertainty representations of the one-sided 95% upper limit for the probability of occurrence of a severe consequence accident.

One-sided upper limit	95%
Hybrid approach	$(3.78 \cdot 10^{-8}, \ 1.95 \cdot 10^{-7})$
Purely probabilistic approach	$1.09 \cdot 10^{-7}$
Purely possibilistic approach	$(3.84 \cdot 10^{-9}, \ 4.00 \cdot 10^{-7})$

characterized by the fact that for values of u lower than $2 \cdot 10^{-10}$ the plausibility $Pl([0, \ u))$ obtained by the purely possibilistic approach is lower than the plausibility obtained by the hybrid approach. This effect stems from the transformation of the lognormal probability density functions into triangular possibility distributions that completely neglect the possibility to have probability values lower than the 5th percentile or higher than the 95th percentile of the original lognormal probability distributions. This approximation serves only the purpose of the comparison but may be considered critical because it is not conservative in the sense of risk analysis.

10.7.2   Comparison of the results for the probability of occurrence of a severe consequence accident

In order to interpret the results obtained by applying the different uncertainty propagation methods with respect to the probability of occurrence of a severe consequence accident, credibility intervals are taken from the belief and plausibility measures $Bel([0, \ u))$ and $Pl([0, \ u))$. For example, in Table 10.3 it is shown that the one-sided 95% upper limit obtained by the hybrid approach is within $(3.78 \cdot 10^{-8}, \ 1.95 \cdot 10^{-7})$. Notice that the upper bound $(1.95 \cdot 10^{-7})$ is only slightly larger than the one-sided upper limit obtained by the probabilistic approach $(1.09 \cdot 10^{-7})$, whereas the interval obtained by the purely possibilistic approach is much larger $3.84 \cdot 10^{-9}, 4.00 \cdot 10^{-7}$.

With respect to the hybrid approach, it is remarkable that the size of the interval within which the upper limit can be found results only from the epistemic uncertainty on knowledge of the HED events. Thus, if the risk analyst were interested in reducing the imprecision on the estimation of the one-sided upper limit, he or she should try to reduce the imprecision regarding the characterization of the probability of occurrence of the HED events.

11

Uncertainty representation and propagation in the evaluation of the consequences of industrial activity

In this chapter we consider the problem of uncertainty representation and propagation in the analysis of the consequences of undesirable events. The application is in part based on Ripamonti *et al* (2012) and Ripamonti *et al.* (2013).

11.1 Evaluation of the consequences of undesirable events

One of the fundamental questions of any risk analysis concerns the evaluation of the consequences of undesirable events associated with the project or activity under investigation. This requires the identification, description, and assessment of all direct and indirect effects, such as those on human beings, fauna and flora, soil, water, air, climate, landscape, material assets.

In this respect, the current European Union regulations require to provide an Environmental Impact Assessment (EIA) for all projects which are likely to have a significant impact on the environment. In this chapter, we consider a new facility responsible for the atmospheric dispersion of pollutants. For this kind of project, an EIA consists of the following three main steps:

Uncertainty in Risk Assessment: The Representation and Treatment of Uncertainties by Probabilistic and Non-Probabilistic Methods, First Edition. Terje Aven, Piero Baraldi, Roger Flage and Enrico Zio.
© 2014 John Wiley & Sons, Ltd. Published 2014 by John Wiley & Sons, Ltd.

1. Characterization of the source: that is, the estimation of the atmospheric emissions resulting from the operation of the plant, in terms of nature and quantities of pollutants.

2. Estimation of the atmospheric pollutant concentration levels in the area close to the new installation using proper atmospheric dispersion models.

3. Assessment of the pollutant concentration levels at the receptor points, resulting from the superposition of the estimated concentration on the existing background concentration levels.

The uncertainties to which these three steps of the EIA procedure are subject may originate from the randomness of the physical processes involved or from a lack of precise knowledge of the model parameters used to describe them.

11.2   Case study

In this case study we consider the project of a new waste gasification plant. The final objective is an assessment of the dioxin/furans (PCDD/Fs) long-term concentration caused by flue gas emissions from the stack in the surroundings of the plant.

The PCDD/Fs air concentration on an annual average basis, C_{air}, can be computed by multiplying an emission term, Q, which quantifies the mass flow rate of PCDD/Fs released into the atmosphere, by an atmospheric dispersion factor DF, which quantifies the ambient concentration per unit of mass flow rate:

$$C_{air} = Q \cdot DF. \tag{11.1}$$

The measurement units are: C_{air}, fg m^{-3}; Q, ng s^{-1}; and DF, fg m^{-3}/ng s^{-1}. Since the atmospheric dilution of the emission occurs at different extents depending on the distance from the stack, C_{air} and DF are actually space dependent and take different values, $C_{air(x,y)}$ and $DF(x, y)$, at the receptor points located in the study domain, usually described by a Cartesian grid centered on the stack. The emitted PCDD/Fs mass flow rate Q in (11.1) is computed on the basis of the plant's daily throughput of waste, T (Mgwaste day^{-1}), the PCDD/Fs concentration in the emitted flue gas, C_D (ng m^{-3}), and the specific gas production V_F (m^3 Mgwaste^{-1}):

$$Q = \frac{T \cdot C_D \cdot V_F}{3600 \cdot 24}, \tag{11.2}$$

with the PCDD/Fs concentration being expressed in terms of equivalent toxicity mass per unit gas volume at normal conditions (0 °C, 101.3 kPa) and, coherently, the specific gas production referring to the same temperature and pressure conditions (Van den Berg et al., 1998; US EPA, 2005).

The dispersion term DF in (11.1) has been computed by simulating the atmospheric transport and dispersion of the emitted PCDD/Fs in the surroundings of the plant by means of the ISCST3 model (Industrial Source Complex – Short-Term version 3 model) (US EPA, 1995). Local meteorological conditions (wind speed and direction, ambient air temperature, atmospheric stability), source features (stack height, flue gas speed and temperature, pollutant mass flow rate), and geographical features of the area (terrain, elevation, land use) are the basic input data for the dispersion model. The output of the ISCST3 model is a two-dimensional field of the contribution to the ground-level concentration data in the study domain around the plant. Since in this model the output concentrations are linearly proportional to the emitted flow rate, the dispersion term DF can be evaluated by running the model for a unit mass flow rate ($1 \, \text{ng s}^{-1}$). The DF values to be used in (11.1) for estimating the PCDD/Fs annual average concentration have been obtained for 1681 nodes of a Cartesian grid centered on the plant (250 m cell spacing) during one-year-long model simulations, based on the hourly time series of locally measured meteorological data.

The input parameters of the emission model (see (11.2)) have been characterized as follows:

- The plant's daily throughput T is a constant parameter whose value is assigned during project design; for the plant in this case study, T has been set to a value of 900 Mgwaste day^{-1}.

- The PCDD/Fs concentration C_D is an uncertain parameter whose value varies during normal gasification operation due to fluctuations in the process parameters and to the heterogeneous and variable composition of the waste.

- The specific gas production V_F is an uncertain parameter. Although its value is usually set in the plant design phase, V_F can present variations during plant operation caused by fluctuations in the energy content of the waste, as a consequence of its heterogeneous and variable composition.

Application 11.1   (in a nutshell)

Input uncertain quantities:

- PCDD/Fs concentration in the emitted flue gas, C_D

- specific gas production, V_F

- atmospheric dispersion factor, DF.

Output quantity:

- the PCDD/Fs air concentration on an annual average basis, C_{air}.

Model:

- $C_{air} = Q \cdot DF$ with $Q = \dfrac{T \cdot C_D \cdot V_F}{3600 \cdot 24}$.

Type of uncertainty on the input quantities:

- epistemic on C_D and DF

- aleatory on C_D and DF.

Uncertainty propagation setting:

- level 1.

11.3 Uncertainty representation

Due to the rather limited applications of the waste gasification process, large data sets containing C_D and V_F values collected during the operation of similar plants are not available. However, different studies have recently investigated pollutant emission from waste gasification plants: see Klein (2002), Yamada, Shimizu, and Miyoshi (2004), Porteous (2005), BioEnergy Producers Association (2009), and Arena (2012). From these studies, 35 values of C_D and 4 values of V_F were derived. The availability of a statistically significant set of C_D average values makes it possible to represent the uncertainty on this model parameter by means of a probability distribution. A Kolmogorov–Smirnov test considering lognormal, Weibull, beta, and logistic distributions was performed in order to properly choose the probability distribution best representing the 35 C_D data. Figure 11.1 shows the selected beta PDF of parameters $\alpha = 0.36$ and $\beta = 1.32$ and maximum value $0.07\,\mathrm{ng\,m^{-3}}$.

Conversely, for V_F the available information (four literature values and the case study design value) is very scarce. This motivated the adoption of a possibility distribution to describe the epistemic uncertainty on this parameter. In particular, a triangular distribution with range $[3360,\ 6670]\ (\mathrm{m^3\,Mgwaste^{-1}})$, corresponding to the minimum and maximum of the four available literature

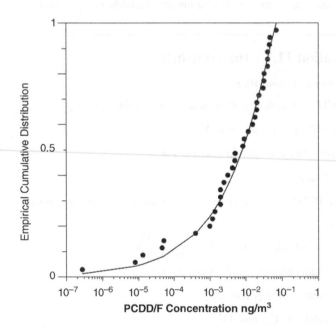

Figure 11.1   Empirical cumulative distribution (•) of the data used to estimate the PCDD/Fs concentration, C_D, and beta distribution used to fit the data (based on Ripamonti et al., 2013).

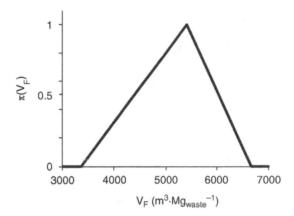

Figure 11.2   Possibility distribution of the specific gas production, V_F *(based on Ripamonti* et al., *2013).*

values, and most likely value set equal to the case study design value (Figure 11.2), was used.

Furthermore, model predictions of the *DF* values in (11.2) are uncertain quantities subject to:

- the natural variability in the input parameters (such as meteorological variables and source features)

- measurement errors

- modeling errors due to the difficulties in capturing atmospheric behavior (Sax and Isakov, 2003; Rao, 2005).

Neglecting model uncertainties for simplicity, the *DF* values are mainly subject to the variability in the meteorological input parameters. Here, the uncertainty in *DF* has been estimated indirectly, by taking into account 10 different years of local meteorological data and by separately running the dispersion model for each input data set. Thus, 10 yearly *DF* values have been estimated for each grid node in the computation domain.

For illustrative purposes, only one grid point has been considered for the uncertainty propagation, chosen as the receptor point most impacted by the plant emission (x_M, y_M); that is, the grid node characterized by the highest average of the 10 yearly *DF* values.

Also in this case, due to the small data set available, a possibility distribution has been used to describe the uncertainty on $DF(x_M, y_M)$. A trapezoidal possibility distribution based on analyst judgments was employed (Figure 11.3). The minimum value of its support is equal to 0, its core ranges between $5.69 \cdot 10^{-3}$ and $4.19 \cdot 10^{-2}$ fg m^{-3}/ng s^{-1}, corresponding to the minimum and maximum values of the 10 estimated *DF* values, and the maximum value of its support is equal to

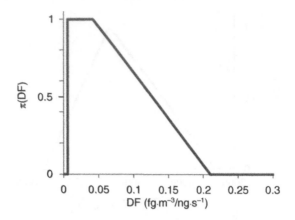

Figure 11.3   Trapezoidal possibility distribution used to represent the uncertainty on the dispersion term DF *(based on Ripamonti* et al., *2013).*

$0.21\,\mathrm{fg\,m^{-3}/ng\,s^{-1}}$, which is the value of DF obtained assuming the worst day conditions observed during the 10-year period for atmospheric dispersion.

11.4   Uncertainty propagation

The hybrid probabilistic–possibilistic uncertainty propagation method illustrated in Section 6.1.3 has been used to propagate the uncertainty from the model input quantities, C_D, V_F, and $DF(x_M, y_M)$, to the model output, C_{air}. In order to perform the propagation, we have used $M = 1000$ Monte Carlo realizations of the parameter C_D and 21 α-cut values (range 0–1, step 0.05) for the possibilistic variables V_F and $DF(x_M, y_M)$. These values of M and the number of α-cuts correspond to those used in the case study of Chapter 10.

11.5   Results

Within the EIA procedure, results are typically communicated in the form of percentiles of the distribution of the PCDD/Fs annual average at the most impacted receptor point in the area. To this end, the information obtained from the uncertainty propagation method, which was represented in the previous application by the belief and plausibility curves of the set $[0, u]$, has been collected in Table 11.1. Within the hybrid probabilistic–possibilistic framework, the 95th percentile of the PCDD/Fs annual average at the most impacted receptor point in the area is an uncertain quantity itself, whose true value lies in the interval $[1.15 \cdot 10^{-2},\ 5.65 \cdot 10^{-1}]\,\mathrm{fg\,m^{-3}}$, where the two limiting values are the 95th percentile of the $Bel([0, u))$ and $Pl([0, u))$ distributions.

Table 11.1   Estimated PCDD/Fs ambient air concentration $C_{air}(x_M, \ y_M)$ (fg m^{-3}): comparison of percentiles obtained by the purely probabilistic method and the hybrid probabilistic–possibilistic method.

Method	Percentile		
	0.5	0.75	0.95
Purely probabilistic	$2.4 \cdot 10^{-2}$	$8.5 \cdot 10^{-2}$	$2.8 \cdot 10^{-1}$
Hybrid probabilistic– possibilistic	$[9.6 \cdot 10^{-4}, \\ 7.50 \cdot 10^{-?}]$	$[3.7 \cdot 10^{-3}, \\ 2.30 \cdot 10^{-1}]$	$[1.2 \cdot 10^{-2}, \\ 5.7 \cdot 10^{-1}]$

11.6   Comparison of the results to those obtained using a purely probabilistic approach

In order to apply a purely probabilistic approach, the possibility distributions used to represent the uncertainty on the quantities V_F and $DF(x_M, y_M)$ have been transformed into probability distributions according to (10.3). Then, the Monte Carlo procedure described in Section 6.1.1, using $M = 10\,000$ samples of each of the three uncertain input quantities from the corresponding probability density functions, was applied. Table 11.1 gives the values of different percentiles for both the purely probabilistic and the hybrid probabilistic–possibilistic methods: the values of the former are always in the range of values of the latter; and the distance between $Bel([0, u))$ and $Pl([0, u))$ increases from lower to higher percentiles. Notice that for any considered percentile the gap between the plausibility and the belief measures is in the range of one/two orders of magnitude. Thus, we can conclude that the uncertainty on the V_F and DF parameters significantly affects the estimated ground-level concentrations. In this respect, the capability of the hybrid probabilistic–possibilistic method to separately process the contribution of the different uncertainties gives rise to results that clearly show the effects of the lack of knowledge on the input parameters. This is a desirable quality leading to more informative and transparent outputs, also in light of the subsequent calculations for health risk assessment.

It is also worth noticing that the concentration estimates at the selected receptor point for the most and least favorable years for atmospheric dispersion obtained by means of the traditional deterministic approach are 0.03 fg m^{-3} and 0.24 fg m^{-3}, respectively, thus varying by almost one order of magnitude. This means that the traditional deterministic approach based on just one single-year simulation which is commonly applied can lead to either too precautionary or non-conservative estimates based on the arbitrary choice of the analyst about the reference year to be used in the model simulation. In this respect, a preliminary assessment of the representativeness of the year chosen for the weather model simulation or multi-year modeling is recommended for deterministic approach calculations.

Finally, notice that both the purely probabilistic method and hybrid probabilistic–possibilistic method have been applied under the simplifying assumption of independence of the parameters C_D and V_F, although they are expected to be somehow correlated.

12

Uncertainty representation and propagation in the risk assessment of a process plant

The presentation in this chapter is largely taken from Aven and Kvaløy (2002).

12.1 Introduction

In this chapter we use a simple risk analysis example as a starting point for a discussion on how to consider a Bayesian approach to risk analysis and to search for practical solutions. We primarily have in mind standard situations where the risk analysis is used in a decision-making context, for example, in the planning phase of a project, and where we have some relevant background information.

12.2 Case description

We consider a process plant on an offshore oil and gas platform. As part of the risk analysis of the plant, a separate study is to be carried out of the risk associated with the operation of a control room placed in the compressor module. Two people operate the control room. The purpose of the study is to assess the risk to the operators as a result of possible fires and explosions in the module, and to evaluate the effect of implementing risk-reducing measures. Based on this study, a decision will be made on whether to move the control room out of the module or to implement some other risk-reducing measures. The risk is currently considered to be too high, but management is not sure what the best overall arrangement is, taking into account both safety and economy. It is

Uncertainty in Risk Assessment: The Representation and Treatment of Uncertainties by Probabilistic and Non-Probabilistic Methods, First Edition. Terje Aven, Piero Baraldi, Roger Flage and Enrico Zio.
© 2014 John Wiley & Sons, Ltd. Published 2014 by John Wiley & Sons, Ltd.

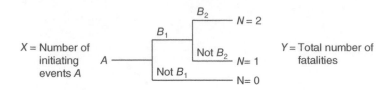

Figure 12.1   Event tree example (based on Aven and Kvaløy, 2002).

therefore decided to conduct a risk analysis to support the decision making. To simplify matters, suppose the analysis is based on an event tree as shown in Figure 12.1. The event tree models the possible occurrence of gas leakages in the process plant during a period of time, say one year. The number of gas leakages, referred to as the initiating events, is denoted by X. If an initiating event A occurs, it leads to N number of fatalities, where $N = 2$ if events B_1 and B_2 occur, $N = 1$ if events B_1 and not B_2 occur, and $N = 0$ if event B_1 does not occur. We may think of event B_1 as representing ignition of the gas and B_2 as explosion. The total number of fatalities is denoted by Y.

In the following we will present first a level 2 and, then, a level 1 analysis, in Sections 12.3 and 12.4 respectively.

12.3   The "textbook" Bayesian approach (level 2 analysis)

In this section we approach the problem presented in Section 12.2 using the prevailing "textbook" Bayesian thinking. We first present the calculations that will typically be performed and then we discuss various interpretations of what the calculations really mean.

Let K denote the background knowledge we have about the problem, that is, all sorts of information which is relevant to the problem. This information might come from various sources, for example, general information from similar situations, more or less relevant historical data from similar situations, expert judgments, and so on. The entire analysis is conditional on this background knowledge.

We begin by specifying a probability model for the problem, consisting of probability distributions for unknown observable quantities as well as for parameters. First consider the number X of initiating events. A typical choice is to use a Poisson distribution with parameter λ, where the probability mass function is given by

$$p(x \mid \lambda) = P(X = x \mid \lambda) = \frac{\lambda^x}{x!} e^{-\lambda}. \tag{12.1}$$

A prior subjective probability distribution $h(\lambda \mid K)$ for λ is then specified, using for instance a gamma distribution. The gamma distribution is a mathematically convenient choice as a prior distribution in this case, since it is a so-called conjugate distribution. A prior distribution is called conjugate if it leads to a posterior distribution (posterior distributions are discussed later) in the same distribution class (in this case gamma), see Bernardo and Smith (1994) for more details.

Next, we define $\theta_1 = P(B_1 \mid A)$ and $\theta_2 = P(B_2 \mid A)$. Then

$$P(N = 2 \mid A, \theta_1, \theta_2, K) = P(N = 2 \mid A, \theta_1, \theta_2) = \theta_1 \theta_2,$$
$$P(N = 1 \mid A, \theta_1, \theta_2, K) = P(N = 1 \mid A, \theta_1, \theta_2) = \theta_1 (1 - \theta_2), \qquad (12.2)$$
$$P(N = 0 \mid A, \theta_1, \theta_2, K) = P(N = 0 \mid A, \theta_1, \theta_2) = 1 - \theta_1.$$

Prior subjective probability distributions $h(\theta_1 \mid K)$ and $h(\theta_2 \mid K)$ are specified, using for instance beta distributions. Note that, more generally, a joint prior distribution $h(\lambda, \theta_1, \theta_2 \mid K)$ of all parameters could be specified, but it is most common to use independent prior distributions, that is,

$$h(\lambda, \theta_1, \theta_2 \mid K) = h(\lambda \mid K) h(\theta_1 \mid K) h(\theta_2 \mid K).$$

A conditional subjective probability distribution of the total number of fatalities can be specified as

$$p(y \mid x, \theta_1, \theta_2) = P(Y = y \mid X = x, \lambda, \theta_1, \theta_2, K) = P(Y = y \mid X = x, \theta_1, \theta_2)$$
$$= P\left(\sum_{i=1}^{x} N_i = y \mid X = x, \theta_1, \theta_2 \right), \qquad (12.3)$$

where N_i is the number of fatalities in the ith occurrence of the initiating event. Calculating this probability is in principle straightforward, but for large x and y it is tedious. The simplest cases are

$$p(0 \mid x, \theta_1, \theta_2) = (1 - \theta_1)^x,$$
$$p(1 \mid x, \theta_1, \theta_2) = x(1 - \theta_1)^{x-1} \theta_1 (1 - \theta_2), \qquad (12.4)$$
$$p(2 \mid x, \theta_1, \theta_2) = x(1 - \theta_1)^{x-1} \theta_1 \theta_2 + \frac{x(x-1)}{2} (1 - \theta_1)^{x-2} \theta_1^2 (1 - \theta_2)^2.$$

A conditional joint distribution of X and Y can now be specified as

$$p(x, y \mid \lambda, \theta_1, \theta_2, K) = p(x, y \mid \lambda, \theta_1, \theta_2) = p(y \mid x, \theta_1, \theta_2) p(x \mid \lambda), \qquad (12.5)$$

and unconditionally

$$p(x, y \mid K) = \int_{\theta_1} \int_{\theta_2} \int_{\lambda} p(y \mid x, \theta_1, \theta_2) p(x \mid \lambda) h(\lambda, \theta_1, \theta_2 \mid K) \, d\lambda \, d\theta_2 \, d\theta_1. \qquad (12.6)$$

Finally, the unconditional distribution of the total number of fatalities becomes

$$p(y \mid K) = \sum_{x} \int_{\theta_1} \int_{\theta_2} \int_{\lambda} p(y \mid x, \theta_1, \theta_2) p(x \mid \lambda) h(\lambda, \theta_1, \theta_2 \mid K) \, d\lambda \, d\theta_2 \, d\theta_1. \qquad (12.7)$$

In the above set-up, $p(x \mid \lambda)$, $p(y \mid x, \theta_1, \theta_2)$, and $p(x, y \mid \lambda, \theta_1, \theta_2, K)$, as well as θ_1 and θ_2, are to be understood as chances, that is, as limits of infinite exchangeable sequences (see Section 2.4); or alternatively, in a probability of frequency set-up, as frequentist probabilities. On the other hand, $p(y \mid K)$ is a subjective probability reflecting both the (aleatory) uncertainty captured by the underlying chances and the epistemic uncertainty related to the chance values.

If additional information in the form of observations $z = (x_1, y_1, \ldots, x_n, y_n)$ of X and Y (not previously included in K) becomes available, the so-called likelihood function is given as

$$L(\lambda, \theta_1, \theta_2 \,|\, z, K) = \prod_{i=1}^{n} p(x_i, y_i \,|\, \lambda, \theta_1, \theta_2, K). \tag{12.8}$$

Alternatively, if data on the number of fatalities for each initiating event is available, that is, $z = (n_1, \ldots, n_x)$, then the likelihood function becomes

$$L(\lambda, \theta_1, \theta_2 \,|\, z, K) = \prod_{i=1}^{x} P(N_i = n_i \,|\, \theta_1, \theta_2, A) P(X = x \,|\, \lambda). \tag{12.9}$$

The prior distributions can then be updated to posterior distributions by using Bayes' theorem. For instance,

$$h(\lambda \,|\, z, K) \propto L(\lambda, \theta_1, \theta_2 \,|\, z, K) h(\lambda \,|\, K), \tag{12.10}$$

where the constant of proportionality ensures that the posterior distribution is a proper distribution, that is, it integrates to 1. The posterior distributions $h(\theta_1 \,|\, z, K)$ and $h(\theta_2 \,|\, z, K)$ are calculated similarly.

Application 12.1   (in a nutshell (level 2))

Input uncertain quantities:

- number of initiating events, X
- outcome of branching events, B_1 and B_2
- Poisson distribution rate parameter, λ
- event tree branching event chances, θ_1 and θ_2.

Output quantity:

- number of fatalities, Y.

Model:

- see (12.7).

The assessment concerns computation of the probability distribution of the number of fatalities, Y.

Type of uncertainty on the input quantities:

- aleatory on X, B_1 and B_2
- epistemic on λ, θ_1 and θ_2.

Uncertainty propagation setting:

- level 2.

The above uncertainty description is an example of a level 2 analysis. In the following section we consider a level 1 analysis which is not based on the introduction of chances, but instead on the use of subjective probabilities directly describing uncertainty about the observable unknown quantities involved, that is, about X, B_1, B_2, and Y.

12.4 An alternative approach based on subjective probabilities (level 1 analysis)

Consider again the event tree example. The focus is on the quantity Y, the number of fatalities. To predict this number and to assess uncertainties, we develop a deterministic model, which is the event tree model shown in Figure 12.1. Given the model, the remaining uncertainties are related to the observable quantities and events, that is, related to X, A and B. The next step is, then, to assess the uncertainties of these quantities, starting with X.

We would like to predict X and assess uncertainties. How should this be done? Suppose that data from situations "similar" to the one analyzed is available, and assume, for the sake of simplicity that the data is of the form $x_1, x_2, \ldots, x_n$, where x_i is the number of initiating events during one year. This data is considered relevant for the situation being studied.

The data allows for a prediction simply by using the mean $\bar{x}$ of the observations $x_1, x_2, \ldots, x_n$. But what about the uncertainty in this prediction? How should we express the uncertainty related to X and the prediction of X? Suppose the observations $x_1, x_2, \ldots, x_n$ are 1, 1, 2, 0, 1, so that $n = 5$ and the observed mean is equal to 1. In this case, we have rather strong background information, and we suggest using the Poisson distribution with mean 1 as the uncertainty distribution of X. How can this uncertainty distribution be "justified?" Well, if this distribution reflects our uncertainty about X, it is justified and there is nothing more to say. This is a subjective probability distribution and there is no need for further justification. But is a Poisson distribution with mean 1 "reasonable," given the background information? We note that this distribution has a variance not greater than 1. By using this distribution, more than 99% of the mass is on values less than or equal to 4.

Adopting the prevailing Bayesian thinking, as outlined above, using the Poisson distribution with mean 1, means that we have no uncertainty about the parameter λ, which is interpreted as the long-run average number of failures when considering an infinite number of exchangeable random quantities representing similar systems to the one being analyzed. According to Bayesian theory, ignoring the uncertainty about λ gives misleading, over-precise inference statements about X (Bernardo and Smith, 1994, p. 483). This reasoning is of course valid if we work within the standard Bayesian setting, considering an infinite number of exchangeable random quantities. In our case, however, we just have one X, so what do we gain by making a reference to limiting quantities of a sequence of similar hypothetical Xs? The point is that, given the observations $x_1, x_2, \ldots, x_5$, the choice of the Poisson distribution with mean 1 is in fact reasonable. Consider the following argumentation.

Suppose that we divide the year $[0, T]$ into time periods of length T/k, where k is for example 1000. Then we may ignore the possibility of having two events occurring in one time period, and we assign an event probability of $1/k$ for the first time period, as we predict one event in the whole interval $[0, T]$. Suppose that we have observations related to $i - 1$ time periods. Then for the next time period we could take these observations into account as using independence means and ignoring available information. A natural way of balancing the prior information and the observations is to assign an event probability of $(d_i + 1 \times n)/((i - 1) + nk)$, where d_i is equal to the total number of events that occurred in $[0, T(i - 1)/k]$; that is, we assign a probability which is equal to the total number of events that occurred per unit of time. It turns out that this assignment process gives an approximate Poisson distribution for X. This can be shown for example by using Monte Carlo simulation. Figure 12.2 shows a histogram resulting from a Monte Carlo simulation of the set-up above using 10^4 simulation runs, as well as the Poisson distribution with mean 1.

The Poisson distribution is justified as long as the background information dominates the uncertainty assessment of the number of events occurring in a time period. Thus from a practical point of view, there is no problem in using the Poisson distribution with mean 1. The above reasoning provides a "justification" of the Poisson distribution, even with not more than one or two years of observations.

Alternatively, the Poisson approximation follows by studying the predictive distribution of X in a full Bayesian analysis, assuming that $x_1, x_2, \ldots, x_5$ are observations coming from a Poisson distribution, given the mean λ and using a suitable (e.g., a non-informative) prior distribution on λ. Restricting attention to observable quantities only, a procedure as specified by Barlow (1998, Chapter 3), can

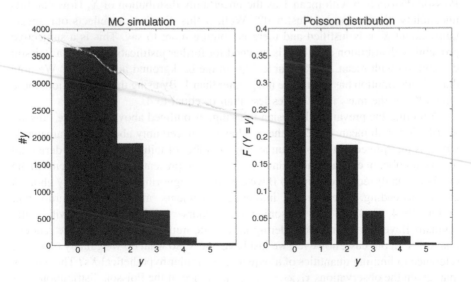

Figure 12.2 Monte Carlo simulation histogram and Poisson distribution.

be used. This procedure, in which the multinomial distribution is used to establish the Poisson distribution, is based on exact calculation of the conditional probability distribution of the number of events in subintervals, given the observed number of events for the whole interval.

Note that for the direct assignment procedure using the k time periods, the observations $x_1, x_2, \ldots, x_5$ are considered part of the background information, meaning that this procedure does not involve any modeling of this data. In contrast, the more standard Bayesian approach requires that we model $x_1, x_2, \ldots, x_5$ as observations coming from a Poisson distribution, given the mean l.

We conclude that a Poisson distribution with mean 1 can be used to describe the analyst's uncertainty with respect to X in this case. The background information is sufficiently strong. We now turn to how to assess the uncertainties for B_1 and B_2. For these events we just need to assign two probabilities, namely, $\theta_1 = P(B_1 \mid A, K)$, expressing our uncertainty related to the occurrence of ignition, and $\theta_2 = P(B_2 \mid B_1, K)$, expressing our uncertainty regarding an explosion given the ignition. The basis for the probability assignments would be "hard" data and expert opinions. These probabilities are not "true underlying probabilities" or limiting frequencies of 0-1 events; they just represent our subjective uncertainties regarding the observable events B_1 and B_2, expressed as probabilities. This is different from the common approach in Bayesian analyses where prior distributions expressing the uncertainties regarding the "true values" of θ_1 and θ_2 are usually specified. Why introduce such hypothetical limiting quantities and associated prior distributions when we can easily assess our uncertainties regarding what would happen by the single numbers θ_1 and θ_2?

What now remains is to use probability calculus to calculate the predictive uncertainty distribution for the total number of fatalities Y. This distribution is now straightforwardly calculated as

$$p(y \mid K) = \sum_x p(y \mid x, \theta_1, \theta_2) p(x \mid K), \qquad (12.11)$$

where $p(x \mid K)$ is the uncertainty distribution of X. Comparing (12.11) to (12.7), we see that the computational simplifications by not having prior distributions on θ_1 and θ_2 are considerable.

So the end product of the analysis is simply the predictive uncertainty distribution $p(y \mid K)$ in 12.11, expressing our uncertainty regarding the future value of Y. There are no further "uncertainties of uncertainties." The uncertainty distribution regarding the "top-level" quantity, here Y, is calculated by first focusing on the observable quantities on a more "detailed level," in this case X, and B_1 and B_2, establishing uncertainty distributions for these, and then using probability calculus to propagate this into an uncertainty distribution for the top-level quantity Y.

The issue is then, in a real case, which approach should be adopted: the level 1 or the level 2 analysis? To answer this question, the key quantities of interest need to be clarified. If it is clear that the quantities are frequentist probabilities, the probability

---

Application 12.2   (in a nutshell (level 1))

Input uncertain quantities:

- number of initiating events, X
- outcome of branching events, B_1 and B_2.

Output quantity:

- number of fatalities, Y.

Model:

- See Eq. (12.11)

The assessment concerns computation of the probability distribution of the number of fatalities, Y.

Type of uncertainty on the input quantities:

- epistemic on X, B_1, and B_2.

Uncertainty propagation setting:

- level 1.

---

model approach – that is, the probability of frequency approach (level 2 analysis) – should be adopted. If it is not clear what the key quantities of interest are, the following question needs to be asked: Is it important to have at hand a framework where new information can be systematically incorporated? If the answer is yes, the probability of frequency should be adopted, provided that frequentist probabilities can be justified. In all other cases, the unique event case (level 1 analysis) should be adopted; see Aven (2012).

References – Part III

Ardillon, E. (2010) *SRA into SRA: Structural Reliability Analyses into System Risk Assessment*, Det Norske Veritas, Høvik, Oslo, pp. 81–108.

Arena, U. (2012) Process and technological aspects of municipal solid waste gasification: a review. *Waste Management*, **32**, 625–639.

Aven, T. (2012) On when to base event trees and fault trees on probability models and frequentist probabilities in quantitative risk assessments. *International Journal of Performability Engineering*, **8** (3), 311–320.

Aven, T. and Kvaløy, J.T. (2002) Implementing the Bayesian paradigm in practice. *Reliability Engineering and System Safety*, **78**, 195–201.

Baraldi, P. and Zio, E. (2008) A combined Monte Carlo and possibilistic approach to uncertainty propagation in event tree analysis. *Risk Analysis*, **28**, 1309–1325.

Baraldi, P., Zio, E., and Popescu, I.C. (2008) Predicting the time to failure of a randomly degrading component by a hybrid Monte Carlo and possibilistic method. International Conference on Prognostics and Health Management (PHM 2008).

Baraldi, P., Popescu, I.C., and Zio, E. (2010a) Methods of uncertainty analysis in prognostics. *International Journal of Performability Engineering*, **6**, 303–331.

Baraldi, P., Popescu, I.C., and Zio, E. (2010b) Methods for uncertainty analysis in the reliability assessment of a degrading structure, in *SRA into SRA: Structural Reliability Analyses into System Risk Assessment* (ed. E. Ardillon), Det Norske Veritas, Høvik, Oslo, pp. 81–108.

Baraldi, P., Compare, M., and Zio, E. (2012) Dempster–Shafer Theory of Evidence to handle maintenance models tainted with imprecision. 11th International Probabilistic Safety Assessment and Management Conference and the Annual European Safety and Reliability Conference (PSAM11 ESREL 2012), vol. 1, pp. 61–70.

Baraldi, P., Compare, M., and Zio, E. (2013a) Maintenance policy performance assessment in presence of imprecision based on Dempster-Shafer Theory of Evidence, *Information Sciences*. doi: 10.1016/j.ins.2012.11.00.

Baraldi, P., Compare, M., and Zio, E. (2013b) Uncertainty analysis in degradation modeling for maintenance policy assessment. *Proceedings of the Institution of Mechanical Engineers, Part O: Journal of Risk and Reliability*, **227** (3), 267–278.

Barlow, R.E. (1998) *Engineering Reliability*, SIAM, Philadelphia, PA.

Uncertainty in Risk Assessment: The Representation and Treatment of Uncertainties by Probabilistic and Non-Probabilistic Methods, First Edition. Terje Aven, Piero Baraldi, Roger Flage and Enrico Zio.
© 2014 John Wiley & Sons, Ltd. Published 2014 by John Wiley & Sons, Ltd.

Bernardo, J.M. and Smith, A. (1994) *Bayesian Theory*, John Wiley & Sons, Ltd, Chichester.

Bigerelle, M. and Lost, A. (1999) Bootstrap analysis of FCGR: application to the Paris relationship and to lifetime prediction. *International Journal of Fatigue*, **21**, 299–307.

BioEnergy Producers Association (2009) Evaluation of Emission from Thermal Conversion Technology Processing Municipal Solid Waste and Biomass. Final report from University of California (CE-CERT), Riverside, CA.

Burmaster, D.E. and Hull, D.A. (1997) Using lognormal distributions and lognormal probability plots in probabilistic risk assessments. *Human and Ecological Risk Assessment*, **3** (2), 235–255.

Chatelet, E., Berenguer, C., and Jellouli, O. (2002) Performance assessment of complex maintenance policies using stochastic Petrinets. Proceedings of the European Safety and Reliability Conference (ESREL 2002), Lyon, France, vol. 2, pp. 532–537.

European Union (2000) Directive 2000/76/EC of the European Parliament and of the Council of 4 December 2000 on the incineration of waste. Official Journal, L 332, 28.12.2000.

Henley, E.J. and Kumamoto, H. (1992) *Probabilistic Risk Assessment*, IEEE Press, New York.

Huang, D., Chen, T., and Wang, M.J. (2001) A fuzzy set approach for event tree analysis. *Fuzzy Sets and Systems*, **118**, 153–165.

International Atomic Energy Agency (2007) *IAEA Safety Glossary: Terminology Used in Nuclear Safety and Radiation Protection*, IAEA, Vienna.

Jeong, I.S., Kim, S.J., Song, T.H. *et al.* (2005) Environmental fatigue crack propagation behavior of cast stainless steels under PWR condition. *Key Engineering Materials*, **297–300**, 968–973.

Klein, A. (2002) Gasification: an alternative process for energy recovery and disposal of municipal solid wastes. MS thesis. Columbia University, New York.

Kozin, F. and Bogdanoff, J.L. (1989) Probabilistic models of fatigue crack growth: results and speculations. *Nuclear Engineering and Design*, **115**, 143–171.

Marseguerra, M. and Zio, E. (2000a) System unavailability calculations in biased Monte Carlo simulation: a possible pitfall. *Annals of Nuclear Energy*, **27**, 1589–1605.

Marseguerra, M. and Zio, E. (2000b) Optimizing maintenance and repair policies via a combination of genetic algorithms and Monte Carlo simulation. *Reliability Engineering and System Safety*, **68**, 69–83.

Marseguerra, M. and Zio, E. (2002) *Basics of Monte Carlo Method with Application to System Reliability*, LiLoLe-Verlag, Hagen.

Marseguerra, M., Zio, E., and Martorell, S. (2006), Basics of genetic algorithms optimization for RAMS applications. *Reliability Engineering and System Safety*, **91**, 977–991.

Mustapa, M.S. and Tamin, M.N. (2004) Influence of R-ratio on fatigue crack growth rate behavior of type 316 stainless steel. *Fatigue & Fracture of Engineering Materials & Structures*, **27**, 31–43.

Nuclear Energy Research Center (1995) Nuclear power plant 2 operating living PRA. Report (Draft). Tao Yuan, Taiwan.

Papoulis, A. and Pillai, U. (2002) *Probability, Random Variables, and Stochastic Processes*, 4th edn, McGraw-Hill, New York.

Porteous, A. (2005) Why energy from waste incineration is an essential component of environmentally responsible waste management. *Waste Management*, **25**, 451–459.

Provan, J.W. (1987) *Probabilistic Fracture Mechanics and Reliability*, Martinus Nijhoff, Amsterdam.

Pulkkinen, U. (1991) A stochastic model for wear prediction through condition monitoring, in *Operational Reliability and Systematic Maintenance* (ed. K. Holmberg and A. Folkeson), Elsevier, London, pp. 223–243.

Rao, K.S. (2005) Uncertainty analysis in atmospheric dispersion modelling. *Pure and Applied Geophysics*, **162**, 1893–1917.

Reliability Manual for Liquid Metal Fast Reactor (LMFBR) Safety Programs (1974) General Electric Company International, Rep. SRD 74-113.

Remy, E., Idée, E., Briand, P., and François, R. (2010) Bibliographical review and numerical comparison of statistical estimation methods for the three-parameters Weibull distribution. Proceedings of the European Safety and Reliability Conference (ESREL 2010), Rhodes, Greece, pp. 219–228.

Ripamonti, G., Lonati, G., Baraldi, P. *et al.* (2012) Uncertainty propagation methods in dioxin/furans emission estimation models. Proceedings of the European Safety and Reliability Conference (ESREL 2011), Troyes, France, pp. 2222–2229.

Ripamonti, G., Lonati, G., Baraldi, P. *et al.* (2013) Uncertainty propagation in a model for the estimation of the ground level concentration of dioxin/furans emitted from a waste gasification plant. *Reliability Engineering and System Safety*, in press.

Ritchie, R.O. (1999) Mechanisms of fatigue crack propagation in ductile and brittle solids. *International Journal of Fracture*, **100**, 55–83.

SAFERELNET: Framework document on maintenance management (2006) http://www.mar.ist.utl.pt/saferelnet/overview.asp.

Sax, T. and Isakov, V. (2003) A case study for assessing uncertainty in local-scale regulatory air quality modelling applications. *Atmospheric Environment*, **37**, 3481–3489.

US, EPA (1995) User's Guide for the Industrial 1 Source Complex (ISC3) Dispersion Models. US Environmental Protection Agency report EPA-454/B-95-003b.

US, EPA (2005) Human Health Risk Assessment for Hazardous Waste Combustion Facilities. Report EPA/530/R-05/006, US EPA Office of Solid Wastes.

Van den Berg, M., Birnbaum, L., Bosveld, B.T.C. *et al.* (1998) Toxic Equivalency Factors (TEFs) for PCBs, PCDDs, PCDFs for humans and wildlife. *Environmental Health Perspectives*, **106**, 775–792.

Yager, R.R. (1996) Knowledge-based defuzzification. *Fuzzy Sets and Systems*, **80**, 177–185.

Yamada, S., Shimizu, M., and Miyoshi, F. (2004) Thermoselect Waste Gasification and Reforming Process. JFE technical report, no. 3, pp. 20–24.

Zille, V., Berenguer, C., Grall, A. *et al.* (2007) Modelling and performance assessment of complex maintenance programs for multi-component systems. Proceedings of the 32nd ESReDA Seminar and 1st ESReDA–ESRA Seminar, Alghero, Italy, pp. 127–140.

Zille, V., Despujols, A., Baraldi, P. *et al.* (2009) A framework for the Monte Carlo simulation of degradation and failure processes in the assessment of maintenance programs performance. Proceedings of the European Safety and Reliability Conference (ESREL 2009), Prague, Czech Republic, pp. 653–958.

Zio, E. (2007) *An Introduction to the Basics of Reliability and Risk Analysis*, World Scientific, Singapore.

Zio, E. (2009) Reliability engineering: old problems and new challenges. *Reliability Engineering and System Safety*, **94**, 125–141.

Zio, E. (2012) *The Monte Carlo Simulation Method for System Reliability and Risk Analysis, Springer Series in Reliability Engineering*, Springer Verlag, Berlin.

Zio, E. and Compare, M. (2011) A snapshot on maintenance modeling and applications. *Marine Technology and Engineering*, **2**, 1413–1425.

Zio, E. and Compare, M. (2013) Evaluating maintenance policies by quantitative modeling and analysis. *Reliability Engineering and System Safety*, **109**, 53–65.

Part IV
CONCLUSIONS

13

Conclusions

The representation and characterization of uncertainties in risk assessment is a serious matter, as uncertainties feature strongly in the decision-making process involved in the management of risk. In looking for a general framework for treating uncertainties in risk assessment, we started with the probabilistic treatment of uncertainties, recognizing its merits and limitations, and thus ventured beyond probability to describe uncertainties in a risk assessment context whose setting demands an extension of concepts and methods. This has led us to consider alternative approaches for representing and characterizing uncertainty, including those based on interval probability, possibility theory, and evidence theory. We have made the point, strongly, that extending the framework for uncertainty analysis naturally leads to extending the framework for risk assessment and management. In much of the existing literature on the representation and analysis of uncertainty, risk is defined in relation to probability. For example, using the well-known triplet definition of risk by Kaplan and Garrick (1981) (see also Kaplan, 1997), risk is equal to the triplet (s_i, p_i, c_i), where s_i is the ith scenario, p_i the probability of that scenario, and c_i the consequence of the ith scenario, $i = 1, 2, \ldots, N$, where N is the total number of scenarios. However, if risk is defined through probabilities we need to clarify what probability means. It obviously cannot be a subjective definition, because we seek a general framework that extends beyond such types of probabilities. Hence, probability must refer to a frequentist/propensity concept. However, frequentist probabilities cannot be justified in cases of non-repeatability and therefore cannot serve as a general concept for risk assessment, applicable to all types of uncertainty representations. Consequently, we have to leave the probability-based risk concepts, and extend our considerations to perspectives on risk that are based on uncertainty instead of probability.

One of the most general risk perspectives is the so-called (C, U) risk perspective, introduced in Section 1.1, where risk is understood as the two-dimensional combination of the (severity of the) consequences C of an activity and associated uncertainties U (what will C be?). This perspective is closely linked to some common risk perspectives in social sciences (Rosa, 1998, 2003; Renn, 2005), which state that risk is basically the same as consequences C or events that could lead to C. The definitions of risk are different, but when it comes to the way risk is to be described, there are strong similarities as the C-type perspective also covers consequences and uncertainties.

Also, the knowledge dimension needs to enter the scene when we try to describe or measure risk. A risk description is obtained by specifying the consequences C and using a description (measure) of uncertainty Q (which could be probability or any other measure, where measure is here interpreted in a wide sense). Specifying the events/consequences means identifying a set of events/quantities of interest C' that characterizes the events/consequences C. An example of C' is the number of fatalities. Depending on the principles laid down for specifying C and on the choice of Q, we obtain different perspectives on how to describe/measure risk. As a general description of risk, we can write (C', Q, K), where K is the knowledge that the specification of C' and the assignment Q are based on. Hence, following this definition, there is a sharp distinction between the risk concept per se and how risk is measured or described.

Instead of (C, U), we often write (A, C, U) when we want to focus on hazards/threats/opportunities A. Similarly we write (A', C', Q, K) in place of (C', Q, K) for the risk description. Vulnerability "given A" can, then, be defined as $(C, U \mid A)$ and a vulnerability description covers $(C', Q, K \mid A)$: vulnerability given an event A is basically risk conditional on this event.

We see that such a way of understanding and describing risk allows for all types of uncertainty representations, and it could consequently serve as a basis of a unified perspective for treating uncertainties in risk assessments.

In this book, we have studied alternative ways of representing and treating the uncertainties in a risk assessment context given this broad understanding of risk. We have looked in five principal directions for the uncertainty representations and treatment:

1. Subjective probability

2. Non-probabilistic representations with the interpretation as lower and upper probabilities

3. Non-probabilistic representations with interpretations other than lower and upper probabilities (degree of belief, degree of possibility, etc.)

4. Hybrid combinations of probabilistic and non-probabilistic representations

5. Semi-quantitative approaches.

These directions are not mutually exclusive, because for example direction 4 could be based on a combination of 1 and 2, and 5 could be seen as a special case of 4 since it is based on the combination of a quantitative approach (i.e., direction 1, 2, or 3) and qualitative assessments.

Subjective probability is currently the most common approach for also treating epistemic uncertainty in risk analysis. We have reflected on the position that "probability is perfect" and on the need for an extended framework for risk assessment that reflects the separation that practically exists between analyst and decision maker.

We have argued that we need to see beyond probability to adequately reflect uncertainties in a risk assessment context. The main point raised (see Section 1.5) is the fact that, while probabilities can always be assigned under the subjective probability approach, the origin and amount of information supporting the assignments are not reflected by the numbers produced.

However, how we should see beyond probability is not straightforward. A handful of approaches are available, but they are not easily implemented in practice. More research has to be carried out to bring these alternative approaches to an operative state where they can in fact be used in practice, when needed. The development in this direction should have the clear aim of obtaining a unified perspective (covering concepts, principles, theories, and operative approaches) for the representation and characterization of risk and uncertainty, by linking probability and alternative representations of uncertainty. The present book is to be seen as an attempt to provide a basis for such a work and describe current thinking about these issues.

A framework for risk assessment needs to allow for both qualitative and quantitative approaches. Earlier work has to a large extent been quantitative, but we have underlined that the full scope of the risks and uncertainties cannot be transformed into a mathematical formula, using probabilities or other quantitative measures of uncertainty. Numbers can be generated, but these alone would not serve the purpose of the risk assessment: to reveal and describe the risks and uncertainties. Some qualitative approaches linked to probability exist (see Section 7.5), but similar approaches have not been developed for the alternative quantitative approaches (probability-bound analysis, imprecise probabilities, possibility theory, evidence theory).

Finally, earlier attempts at integration (e.g., hybrid probability and possibility approaches) have been based on the idea that there exists one and only one appropriate representation in a specific case (e.g., possibility representation if the information is poor and subjective probabilities if the information is strong). We believe that the variety of decision-making situations calls for a unified perspective that allows the use of several approaches for representing and characterizing the risk and uncertainties. To inform the decision maker, both subjective probabilities and imprecision intervals may be used, as these approaches could capture different types of information and knowledge important for the decision maker. In addition, qualitative approaches could be incorporated to provide an even more nuanced characterization of the risk and uncertainties.

The "non-probabilistic methods" are also based on a set of premises and assumptions, but not to the same degree as the pure probability-based analyses. Their motivation is that the intervals produced correspond better to the information available. A hybrid probability–possibility analysis may result in an interval $[0.2, 0.6]$ (say) for a subjective probability P. The risk analysts (experts) are not able or willing to precisely assign their probability P. The decision maker may, however, request that the analysts make such assignments – the decision maker would like to be informed of the analysts' degree of belief. The analysts are consulted as experts in the field studied and the decision maker expects them to give a faithful report on the epistemic uncertainties about the unknown quantities addressed. The decision maker knows that these judgments are based on some knowledge and some assumptions, and are subjective in the sense that others could conclude differently, but these judgments are still considered valuable as the analysts have competence in the field being studied. They are trained in probability assignments and the decision maker expects that they will be able to transform their knowledge into probability figures.

Our experience, based on many years of work with risk assessment methodologies and applied risk assessments, is that engineers and risk analysts often struggle with the uncertainty analysis part. We hope that this book can provide some help and guidance. However, the book is not a cookbook for how to conduct uncertainty analysis in a risk assessment context; what is covered are the basic ideas, concepts, and some methods for a set of alternative approaches, as well as overall reflections on how to think when addressing uncertainty in these contexts. The theory, together with the examples presented and discussed in the book, should give the reader a solid basis for these topics and serve well as preparation for carrying out uncertainty analyses in practice.

References – Part IV

Kaplan, S. (1997) The words of risk analysis. *Risk Analysis*, **17**, 407–417.

Kaplan, S. and Garrick, B.J. (1981) On the quantitative definition of risk. *Risk Analysis*, **1** (1), 11–27.

Pidgeon, N., Kasperson, R.E., and Slovic, P. (2003) *The Social Amplification of Risk*, Cambridge University Press, Cambridge.

Renn, O. (2005) *Risk Governance*. White paper no. 1, International Risk Governance Council, Geneva.

Rosa, E.A. (1998) Metatheoretical foundations for post-normal risk. *Journal of Risk Research*, **1**, 15–44.

Rosa, E.A. (2003) The logical structure of the social amplification of risk framework (SARF): metatheoretical foundations and policy implications, in (eds. N. Pidgeon, R.E. Kasperson, and P. Slovic), Cambridge University Press, Cambridge, pp. 47–79.

Uncertainty in Risk Assessment: The Representation and Treatment of Uncertainties by Probabilistic and Non-Probabilistic Methods, First Edition. Terje Aven, Piero Baraldi, Roger Flage and Enrico Zio.
© 2014 John Wiley & Sons, Ltd. Published 2014 by John Wiley & Sons, Ltd.

Appendix A

Operative procedures for the methods of uncertainty propagation

This appendix reports the operative procedures for uncertainty propagation according to some of the methods described in Chapter 6.

A.1 Level 1 hybrid probabilistic–possibilistic framework

We assume that the uncertainty related to the first n input quantities, $X_1, \ldots, X_n$, of the function g is described using the probability distributions $F_1(x_1), \ldots, F_n(x_n)$, whereas the uncertainty related to the remaining $N - n$ quantities $X_{n+1}, \ldots, X_N$ is represented using the possibility distributions $\pi_{n+1}(x_{n+1}), \ldots, \pi_N(x_N)$. The number of repetitions of the Monte Carlo sampling of the input quantities $X_1, \ldots, X_n$ will be referred to as M. The operative steps for the propagation of the hybrid uncertainty information through the function g are (see Figure 6.14)

1. Set $j = 0$.

2. Sample the jth realization $x_1^j, \ldots, x_n^j$ of $X_1, \ldots, X_n$ from the probability distributions $F_1(x_1), \ldots, F_n(x_n)$.

3. Select a value $\alpha \in [0, 1]$ and determine the corresponding α-cuts $(X_{n+1}^\alpha, \ldots, X_N^\alpha)$ of the possibility distributions $\pi_{n+1}(x_{n+1}), \ldots, \pi_N(x_N)$ as intervals of possible values of the possibilistic quantities $(X_{n+1}, \ldots, X_N)$.

Uncertainty in Risk Assessment: The Representation and Treatment of Uncertainties by Probabilistic and Non-Probabilistic Methods, First Edition. Terje Aven, Piero Baraldi, Roger Flage and Enrico Zio.
© 2014 John Wiley & Sons, Ltd. Published 2014 by John Wiley & Sons, Ltd.

4. Calculate the smallest and largest values of $g(x_1^j, \ldots, x_n^j, X_{n+1}^\alpha, \ldots, X_N^\alpha)$, denoted by $\underline{g}_\alpha^j$ and $\overline{g}_\alpha^j$ respectively, considering the fixed values $(x_1^j, \ldots, x_n^j)$ sampled for the random quantities $(X_1, \ldots, X_n)$ and all values of the possibilistic quantities $(X_{n+1}, \ldots, X_N)$ in the α-cuts $(X_{n+1}^\alpha, \ldots, X_N^\alpha)$ of their possibility distributions $\pi_{n+1}(x_{n+1}), \ldots, \pi_N(x_N)$.

5. Take the extreme values $\underline{g}_\alpha^j$ and $\overline{g}_\alpha^j$ found in step 4 as the lower and upper limits of the α-cut of $g(x_1^j, \ldots, x_n^j, X_{n+1}^\alpha, \ldots, X_N^\alpha)$.

6. Return to step 3 and repeat for another α-cut; the possibility distribution π_Z^j of Z is obtained as the collection of values $\underline{g}_\alpha^j$ and $\overline{g}_\alpha^j$ for each α-cut.

7. If $j < M$ set $j = j + 1$ and go back to step 1, otherwise exit from the procedure.

At the end of the procedure, having Monte Carlo-sampled M values of the probabilistic quantities, an ensemble of realizations of possibility distributions is obtained, that is, a set of possibility distributions $(\pi_Z^1, \ldots, \pi_Z^M)$. Then, according to the method described in Section 6.1.3 (Equations (6.16)–(6.19)), the belief $Bel(A)$ and the plausibility $Pl(A)$ for any set A of the output quantity Z can be obtained.

A.2 Level 2 purely probabilistic framework

We assume the presence of N quantities $X_1, \ldots, X_N$ whose uncertainty is character-ized by frequentist probability distributions $F_i(x_i | \theta_i) = P_f(X_i \leq x_i | \Theta_i = \theta_i)$, $i = 1, \ldots, N$, where Θ_i is a vector of the (unknown) parameters of the corresponding probability distribution. The epistemic uncertainty on the parameters Θ_i and the aleatory uncertainty on the input quantities $X_1, \ldots, X_N$ are represented using subjec-tive and frequentist probability distributions, respectively. In particular, we let $h_i(\theta_i)$ denote the subjective probability density function describing the uncertainty of the parameter(s) Θ_i. The number of repetitions of the Monte Carlo sampling of the quantities Θ_i will be referred to as M_e, whereas the number of repetitions of the Monte Carlo sampling of the quantities X_i will be referred as M_a.

The operative steps of the procedure for the propagation of the uncertainty through a function $g(X_1, \ldots, X_N)$ are:

1. Set $j_e = 1$.

2. Sample the j_eth realization $\theta_i^{j_e}$ of the epistemic parameter(s) Θ_i from their respective distributions $h_i(\theta_i)$.

3. Set $j_a = 1$.

4. Sample the j_ath realization $x_1^{j_e, j_a}, \ldots, x_N^{j_e, j_a}$ of the aleatory quantities from their respective distributions $F_i(x_i | \theta_i^{j_e})$, sampled in step 2 for each $i = 1, \ldots, N$.

5. Compute the model output Z corresponding to the realization $x_1^{j_e, j_a}, \ldots, x_N^{j_e, j_a}$ of the input quantities:

$$z^{j_e, j_a} = g(x_1^{j_e, j_a}, \ldots, x_N^{j_e, j_a}).$$

6. If $j_a < M_a$ set $j_a = j_a + 1$ and go back to step 4, otherwise go to step 7.

7. Estimate the cumulative distribution $H^{j_e}(z \mid \theta^{j_e})$ of the model output Z, conditioned by the sampled values $\theta^{j_e} = (\theta_1^{j_e}, \ldots, \theta_N^{j_e})$ of the epistemic quantities, from the obtained z^{j_e, j_a}, $j_a = 1, \ldots, M_a$.

8. If $j_e < M_e$ set $j_e = j_e + 1$ and go back to step 2, otherwise exit the procedure.

The application of this procedure provides a set of a cumulative distributions, $H^{j_e}(z \mid \theta^{j_e})$, $j_e = 1, \ldots, M$, one for each repetition of the outer loop. The interpretation of these distributions is discussed in Section 6.2.

Appendix B

Possibility–probability transformation

The presentation in this appendix is taken from Flage *et al.* (2013). Procedures for the transformation from a possibilistic representation to a probabilistic one, and vice versa, have been suggested; see, for example, Dubois *et al.* (1993). The transformations are not one-to-one, and in going from possibility (probability) to probability (possibility) some information is introduced (lost) in the transformation procedure. However, certain principles can be adopted so that there is minimum loss (introduction) of (artificial) information.

With possibility–probability/probability–possibility transformations, uncertainty propagation can be performed within a single calculus, using Monte Carlo sampling when transforming possibility distributions into probability distributions and fuzzy methods for the converse.

In this appendix we review the possibility–probability transformation method applied in Chapters 8–11 in Part III. Probability–possibility transformation is not considered; we refer to Dubois *et al.* (1993) for an overview of such methods.

We consider the transformation from a possibility distribution into a probability distribution. The transformation is based on given principles and ensures consistency to the extent that there is no violation of the formal rules connecting probability and possibility when possibility and necessity measures are understood as upper and lower probabilities, and so that the transformation is not arbitrary within the constraints of these rules. Nevertheless, as noted by Dubois *et al.* (1993):

> going from a probabilistic representation to a possibilistic representation, some information is lost because we go from point-valued probabilities to interval-valued ones; the converse transformation adds information to

some possibilistic incomplete knowledge. This additional information is always somewhat arbitrary.

Given the interpretation of possibility and necessity measures as upper and lower probabilities, a possibility distribution π induces a family $\mathbf{P}(\pi)$ of probability measures. Since there is not a one-to-one relation between possibility and probability, a transformation of a possibility distribution π into a probability measure P can therefore only ensure that:

- P is a member of $\mathbf{P}(\pi)$; and

- P is selected among the members of $\mathbf{P}(\pi)$ according to some principle (rationale); for example, "minimize the information content of P," in some sense.

We will deal with probability densities. In the following h denotes the probability density associated with a probability measure P.

Different possibility–probability transformations have been suggested in the literature. Dubois *et al.* (1993) argue that the following should be basic principles for such transformations.

I. *The probability–possibility consistency principle*
The family $\mathbf{P}(\pi)$ is formally defined as

$$\mathbf{P}(\pi) = \{P : \forall A \subseteq \Omega, P(A) \le \Pi(A)\},$$

that is, as the set of probability measures P such that for all events A in the space Ω on which π is defined, the probability of A is less than or equal to the possibility of A. As suggested by Dubois *et al.* (1993), it seems natural to require a transformation to select P from $\mathbf{P}(\pi)$. This is referred to as the probability–possibility consistency principle, formulated as

$$P(A) \le \Pi(A), \quad \forall A \subseteq \Omega.$$

II. *Preference preservation*
A possibility distribution π induces a preference ordering on Ω, such that $\pi(x) > \pi(x')$ means that the outcome x is preferred to x'. A transformation should therefore satisfy

$$\pi(x) > \pi(x') \Leftrightarrow p(x) > p(x').$$

The following transformation method is presented by Dubois *et al.* (1993).

i. *Possibility to probability: the principle of insufficient reason*
The principle of insufficient reason specifies that maximum uncertainty on an interval should be described by a uniform probability distribution on that interval. The sampling procedure for transforming a possibility distribution into a probability distribution according to this principle is as follows:

1. Sample a random value α^* in $(0, 1]$ and consider the α-cut level $L_{\alpha^*} = \{x: \pi(x) \geq \alpha^*\}$.

2. Sample x^* at random in L_{α^*}.

In the continuous case, the density h resulting from a transformation of π is given by

$$h(x) = \int_0^{\pi(x)} \frac{d\alpha}{|L_\alpha|}, \tag{B.1}$$

where $|L_\alpha|$ is the length of the α-cut levels of π. To motivate this, note that

$$h(x) = \int_0^1 h(x \mid \alpha)h(\alpha)d\alpha.$$

From step 1 in the sampling procedure above, we have $h(\alpha) = 1$, and from step 2 we have

$$h(x \mid \alpha) = \frac{1}{|L_\alpha|}.$$

For the integration space we note that $h(x|\alpha) = 0$ for $\alpha > \pi(x)$. It should be mentioned that h is the center of gravity of $\mathbf{P}(\pi)$. The transformation in (B.1) applies to upper semi-continuous, unimodal, and support-bounded π.

Other transformation principles are also referred to by Dubois *et al.* (1993), as follows.

ii. *Possibility to probability: the maximum entropy principle*
Select the P in $\mathbf{P}(\pi)$ that maximizes entropy. In general, this transformation violates the preference preservation constraint.

Reference

Dubois, D., Prade, H., and Sandri, S. (1993) On possibility/probability transformations, in R. Lowen and M. Roubens (eds.) *Fuzzy Logic: State of the Art*, Kluwer Academic, Dordrecht, pp. 103–112.

Flage, R., Baraldi, P., Aven, T., and Zio, E. (2013) Probabilistic and possibilistic treatment of epistemic uncertainties in fault tree analysis. *Risk Analysis*, **33** (1), 121–133.

Index

Underlined page numbers denote defined/explained term.

*Uncertainty in Risk Assessment: The Representation and Treatment of Uncertainties by Probabilistic and
Non-Probabilistic Methods*, First Edition. Terje Aven, Piero Baraldi, Roger Flage and Enrico Zio.
© 2014 John Wiley & Sons, Ltd. Published 2014 by John Wiley & Sons, Ltd.

Printed and bound by CPI Group (UK) Ltd, Croydon, CR0 4YY

27/10/2024

14580207-0001